SCIENCE WRITING IN GRECO-ROMAN ANTIQUITY

We access Greek and Roman scientific ideas mainly through those texts which happen to survive. By concentrating only on the ideas conveyed, we may limit our understanding of the meaning of those ideas in their historical context. Through considering the diverse ways in which scientific ideas were communicated, in different types of texts, we can uncover otherwise hidden meanings and more fully comprehend the historical contexts in which those ideas were produced and shared, the aims of the authors and the expectations of ancient readers. Professor Taub explores the rich variety of formats used to discuss scientific, mathematical and technical subjects, from *ca.* 700 BCE to the sixth century CE. Each chapter concentrates on a particular genre – poetry, letter, encyclopaedia, commentary and biography – offering an introduction to Greek and Roman scientific ideas, while using a selection of ancient writings to focus on the ways in which we encounter them.

LIBA TAUB is Director and Curator of the Whipple Museum and Head of the Department of History and Philosophy of Science at the University of Cambridge; she is also a Fellow of Newnham College. She was awarded an Einstein Foundation Visiting Fellowship to work with the Topoi Excellence Cluster (Berlin), and was the recipient of the Joseph H. Hazen Education Prize of the History of Science Society and a University of Cambridge Pilkington Prize for excellence in teaching. She is the author of *Ptolemy's Universe: The Natural Philosophical and Ethical Foundations of Ptolemy's Astronomy* (1993), *Ancient Meteorology* (2003), and *Aetna and the Moon: Explaining Nature in Ancient Greece and Rome* (2008).

KEY THEMES IN ANCIENT HISTORY

EDITORS

P. A. Cartledge
Clare College, Cambridge

P. D. A. Garnsey
Jesus College, Cambridge

Key Themes in Ancient History aims to provide readable, informed and original studies of various basic topics, designed in the first instance for students and teachers of classics and ancient history, but also for those engaged in related disciplines. Each volume is devoted to a general theme in Greek, Roman or, where appropriate, Graeco-Roman history, or to some salient aspect or aspects of it. Besides indicating the state of current research in the relevant area, authors seek to show how the theme is significant for our own as well as ancient culture and society. It is hoped that these original, thematic volumes will encourage and stimulate promising new developments in teaching and research in ancient history.

Other books in the series

Death-Ritual and Social Structure in Classical Antiquity, by Ian Morris 978 0 521 37465 1 (hardback) 978 0 521 37611 2 (paperback)

Literacy and Orality in Ancient Greece, by Rosalind Thomas 978 0 521 37346 3 (hardback) 978 0 521 37742 3 (paperback)

Slavery and Society at Rome, by Keith Bradley 978 0 521 37287 9 (hardback) 978 0 521 37887 1 (paperback)

Law, Violence, and Community in Classical Athens, by David Cohen 978 0 521 38167 3 (hardback) 978 0 521 38837 5 (paperback)

Public Order in Ancient Rome, by Wilfried Nippel 978 0 521 38327 1 (hardback) 978 0 521 38749 1 (paperback)

Friendship in the Classical World, by David Konstan 978 0 521 45402 5 (hardback) 978 0 521 45998 3 (paperback)

Sport and Society in Ancient Greece, by Mark Golden 978 0 521 49698 8 (hardback) 978 0 521 49790 9 (paperback)

Food and Society in Classical Antiquity, by Peter Garnsey 978 0 521 64182 1 (hardback) 978 0 521 64588 1 (paperback)

Banking and Business in the Roman World, by Jean Andreau 978 0 521 38031 7 (hardback) 978 0 521 38932 7 (paperback)

Roman Law in Context, by David Johnston 978 0 521 63046 7 (hardback) 978 0 52163961 3 (paperback)

Religions of the Ancient Greeks, by Simon Price 978 0 521 38201 4 (hardback) 978 0 521 38867 2 (paperback)

Christianity and Roman Society, by Gillian Clark 978 0 521 63310 9 (hardback) 978 0 521 63386 4 (paperback)
Trade in Classical Antiquity, by Neville Morley 978 0 521 63279 9 (hardback) 978 0 521 63416 8 (paperback)
Technology and Culture in Greek and Roman Antiquity, by Serafina Cuomo 978 0 521 81073 9 (hardback) 978 0 521 00903 4 (paperback)
Law and Crime in the Roman World, by Jill Harries 978 0 521 82820 8 (hardback) 978 0 521 53532 8 (paperback)
The Social History of Roman Art, by Peter Stewart 978 0 521 81632 8 (hardback) 978 0 521 01659 9 (paperback)
Ancient Greek Political Thought in Practice, by Paul Cartledge 978 0 521 45455 1 (hardback) 978 0 521 45595 4 (paperback)
Asceticism in the Graeco-Roman World, by Richard Finn OP 978 0 521 86281 3 (hardback) 978 0 521 68154 4 (paperback)
Domestic Space and Social Organisation in Classical Antiquity, by Lisa C. Nevett 978 0 521 78336 1 (hardback) 978 0 521 78945 5 (paperback)
Money in Classical Antiquity, by Sitta von Reden 978 0 521 45337 0 (hardback) 978 0 521 45952 5 (paperback)
Geography in Classical Antiquity, by Daniela Dueck and Kai Brodersen 978 0 521 19788 5 (hardback) 978 0 521 12025 8 (paperback)
Space and Society in the Greek and Roman Worlds, by Michael Scott 978 1 107 00915 8 (hardback) 978 1 107 40150 1 (paperback)
Studying Gender in Classical Antiquity, by Lin Foxhall 978 0 521 55318 6 (hardback) 978 0 521 55739 9 (paperback)
The Ancient Jews from Alexander to Muhammad, by Seth Schwartz 978 1 107 04127 1 (hardback) 978 1 107 66929 1 (paperback)
Language and Society in the Greek and Roman Worlds, by James Clackson 978 0 521 19235 4 (hardback) 978 0 521 14066 9 (paperback)
The Ancient City, by Arjan Zuiderhoek 978 0 521 19835 6 (hardback) 978 0 521 16601 0 (paperback)

SCIENCE WRITING IN GRECO-ROMAN ANTIQUITY

LIBA TAUB
University of Cambridge

CAMBRIDGE
UNIVERSITY PRESS

CAMBRIDGE
UNIVERSITY PRESS

University Printing House, Cambridge CB2 8BS, United Kingdom

Cambridge University Press is part of the University of Cambridge.

It furthers the University's mission by disseminating knowledge in the pursuit of education, learning, and research at the highest international levels of excellence.

www.cambridge.org
Information on this title: www.cambridge.org/9780521130639
DOI: 10.1017/9781139030762

© Liba Taub 2017

This publication is in copyright. Subject to statutory exception and to the provisions of relevant collective licensing agreements, no reproduction of any part may take place without the written permission of Cambridge University Press.

First published 2017

Printed in the United Kingdom by Clays, St Ives plc

A catalogue record for this publication is available from the British Library

Library of Congress Cataloging-in-Publication Data
Names: Taub, Liba Chaia, 1954–
Title: Science writing in Greco-Roman antiquity / Liba Taub, University of Cambridge.
Description: Cambridge: Cambridge University Press, 2017. |
Series: Key themes in ancient history |
Includes bibliographical references and index.
Identifiers: LCCN 2016049752 | ISBN 9780521113700 (hard back) |
ISBN 9780521130639 (paper back)
Subjects: LCSH: Scientific literature – History – To 1500. |
Technical writing – History – To 1500. | Science, Ancient.
Classification: LCC Q225.5.T38 2017 | DDC 509.38–dc23
LC record available at https://lccn.loc.gov/2016049752

ISBN 978-0-521-11370-0 Hardback
ISBN 978-0-521-13063-9 Paperback

Cambridge University Press has no responsibility for the persistence or accuracy of URLs for external or third-party Internet Web sites referred to in this publication and does not guarantee that any content on such Web sites is, or will remain accurate or appropriate.

Contents

List of Illustrations	*page* viii
Preface	ix
Acknowledgements	xiii
Spelling and Abbreviations	xv
Introduction	1
1 Poetry	22
2 Letter	50
3 Encyclopaedia	72
4 Commentary	86
5 Biography	111
Conclusion	130
Appendix A: Arithmetical Epigrams from Book 14 of The Greek Anthology	135
Appendix B: Eratosthenes' Letter to King Ptolemy	144
Bibliographical Essay	149
References	157
Index	181

Illustrations

4.1 Lettered diagram based on the description given by Aristotle in his *Meteorology* 363a34-b7. (The version here is an emendation of that provided by Lee 1952: 187.) *page* 104
4.2 Lettered diagram depicting Philoponus' description of ants tracking a path, following Kupreeva 2012: figure 1, p. 48. 107
A.1 Diagrams based on Cohen and Drabkin 1948: 64, to accompany Eratosthenes' *Letter to King Ptolemy*. 145

Preface

What we today call 'science' has often been described as a Greek invention, even though some historians of science would now argue against this, pointing instead to Mesopotamia as the birthplace of scientific and mathematical practices. Nevertheless, a number of distinguished ancient Greek and Roman thinkers held the view that certain types of explanation had originated in ancient Greece. Aristotle himself (384–322 BCE) credited Thales of Miletus (*fl.* 586 BCE) with having been the first to aim at understanding the original causes of world; in Book 1 (A) of the *Metaphysics*, Aristotle offered a history of explaining the origin of the world, its causes and its composition, naming particular specific individuals and their ideas. This approach to telling the history of scientific thinking – providing an intellectual history concentrating on great individuals and the concepts associated with them – has had a long and fruitful history.

Whilst writing my book *Ancient Meteorology*, I became particularly interested in the ways in which ancient Greek and Roman authors writing on meteorological topics chose to communicate. My study of Greco-Roman works dealing with meteorological phenomena alerted me to the diversity of genres and types of texts used by ancient authors to communicate their ideas and methods for explaining and predicting weather phenomena, texts which include poetry, astrometeorological calendars (known as *parapēgmata*), natural philosophical prose works, letters, question-and-answer texts and commentaries, as well as others. A wide range of styles of writing was deployed; this characteristic of ancient writing on meteorology intrigued me. As I was trained – like many other historians of ancient science – in the tradition of the history of ideas, this diversity took me by surprise, and ignited my curiosity regarding the choices made by ancient authors writing on what we moderns regard as 'science'.

Our knowledge of ancient Greek and Roman science comes to us, largely, through those ancient writings which survive that give an account of scientific and mathematical ideas, methods and practices. This volume explores the rich range and variety of formats used for ancient Greek and Roman writings on scientific, mathematical, medical and technical (here often abbreviated as 'scientific') subjects, paying particular attention to the intellectual and broader cultural contexts in which these works were produced and used. Of course, it is not only the form of communication that is important; what is being communicated is crucial. However, in the past, a primary focus on intellectual history and the history of ideas has seemingly taken for granted that all 'content' can be extracted without considering the medium of communication. I argue that a consideration of the formal features of ancient Greek and Roman writings on scientific topics reveals layers of meaning that cannot be uncovered by concentrating solely on the ideas conveyed. Our understanding of those ideas, as well as of the cultures in which they were produced, communicated, studied and preserved, is enhanced by a deeper engagement with the 'medium' which conveys the message (cf. McLuhan 1964).

Today, readers of ancient scientific, mathematical and technical works do not always come into contact with the form of the original text. Modern readers of such texts, particularly working from translations, may miss the meanings conveyed through formal features, such as metre. A case in point is Lucretius' *De rerum natura*. Readers of the Penguin prose translation of Lucretius' *On the nature of the universe* by R.E. Latham (1994) can be excused if they do not realise that the author was a poet who believed (as he twice notes, 1.921–50 and 4.1–25) that his verse offered an especially appealing version of Epicurean philosophy – in the 'honeyed-cup' of Latin poetry. That Lucretius chose to convey natural-philosophical ideas using epic hexameter is thus lost; the cultural meanings and nuances conveyed by that metre disappear, and our understanding of the ideas contained in the poem is truncated (see Sedley 1998). Perhaps an understanding of this significant limitation of the prose translation explains the publication by Penguin in 2007 of a verse translation by A.E. Stallings, *The Nature of Things*.

A text-based approach to ancient works on scientific and mathematical topics is used here. In drawing attention to the choice of medium used to convey the message, I hope to spark further consideration of the interaction between the two, including the effect of literary conventions associated with particular genres on the presentation of material, as well as its reception by readers. A question springs to mind: Were particular formats or genres used within particular fields of scientific inquiry and, if so, why?

This is an intriguing question to which the answer is not as simple and straightforward as might be imagined. For example, the fundamentals of Epicurean cosmology and physics were communicated in antiquity through several distinct genres, including letters, poetry and biography. This suggests that, for those promoting Epicurean ideas, there was no one favoured medium. Similarly, whilst the formal features of Greek mathematics are distinguishable and were historically extremely influential, mathematics was communicated not only via formal proofs, *à la* Euclid.[1] This suggests that it is not only the relationship between form and content that is important, but also the relationship, for example, between form and function. For what purpose were the ideas being communicated in a specific work? How did the aim of that work relate to the form in which it was composed and presented? Did the purpose(s) for which the text was composed influence the format? It may seem surprising – even counterintuitive to modern sensibilities – that in a number of cases it was a poetic presentation of ideas originally communicated in prose that survived and thrived; as an example, Aratus' poem, the *Phaenomena*, was repeatedly translated from Greek into Latin in antiquity, an indication of its interest and appeal, overshadowing Eudoxus' prose *Phaenomena*, upon which it was based; Eudoxus' text itself did not survive. Did different genres reach different audiences? Convey different messages? Tend to be valued in different ways? These are questions that are difficult to answer in general terms. In recent years, there has been a burgeoning interest in addressing such issues, and more work needs to be done studying specific texts. This volume is an introduction to ancient Greek and Roman scientific ideas and explanations, and focuses on the ways in which we encounter those ideas and explanations, through a selection of those ancient writings that survive.

While some very well-known and important texts by major figures in the history of ancient science are mentioned, what follows is not intended as a comprehensive survey of Greco-Roman scientific, mathematical and technical writings; rather, this book may be read as a series of 'case studies'. In certain cases, texts that are seldom encountered outside of specialist scholarly circles will be central to the discussion, including some by famous thinkers, such as Aristotle and Epicurus. One aim here is to enlarge the readers' acquaintance with unfamiliar examples of Greco-Roman writings about nature and mathematics. Readers are encouraged to read a variety of texts on scientific topics and to be open to considering what might qualify as a scientific or mathematical text. Significantly, the genres examined here were used to convey scientific ideas; however, these

---

[1] See Taub 2013 on genres of Greek mathematics.

same genres were not used only for 'scientific' subjects, but for others as well. This underlines a recurrent theme here: 'science' is part of broader culture, which cannot be simply bracketed off from other cultural expressions. For example, as we will see, 'biographies' of ancient scientists sometimes conveyed strong ethical and religious messages; the formal structure of these corresponds closely, in some instances, to accounts of the life of Jesus conveyed in the gospels.

This volume is concerned with the ways in which ancient Greek and Roman authors communicated scientific ideas and methods through their writings. Following the Introduction, each chapter concentrates on carefully selected texts representative of a particular genre or type of text. Some of the genres discussed (for example, 'biography') were not always well delineated in antiquity;[2] some works incorporate elements of more than one genre. There is not sufficient space available here to provide a full survey of the different genres, including some which were particularly long-lived and are familiar even today, such as the introductory teaching text. The textual formats considered here by no means represent the sum total of options used by ancient Greek and Roman authors writing on scientific, mathematical and technical subjects. Furthermore, there are sometimes complicated issues to confront when describing Greco-Roman scientific and mathematical writings. The process of tackling these issues can provide intriguing insights into the place of these works within the wider cultural contexts of Greco-Roman antiquity. What is presented in this volume is intended to emphasise the value of reading the ancient texts themselves, in order to gain those insights.

---

[2] Carey 2007: 236 points to the 'flexibility of and fluidity between literary forms in living traditions'.

Acknowledgements

It is my pleasure to thank those who have helped in the writing of this book, in a variety of ways. Paul Cartledge, Peter Garnsey and Michael Sharp encouraged me to contribute my work to the Key Themes in Ancient History series, and have been extraordinarily patient and helpful; I am grateful to them for their comments and suggestions throughout. My gratitude is also due to the Einstein Foundation for having supported my work through the award of an Einstein Fellowship at Topoi in Berlin for four years. During my fellowship, I benefitted from the opportunity to meet frequently with colleagues, especially Gerd Graßhoff, Philip van der Eijk and Markus Asper. My Einstein Fellowship also enabled the continuation of a long-running series of workshops devoted to the study of technical texts; I gained a great deal from our collaborators and participants, most notably Jochen Althoff (who also read an earlier version of part of the book), Sabine Föllinger, Harry Hine and Oliver Stoll. Colleagues and friends discussed and read chapters at various stages, offering helpful suggestions, for which I am also grateful. Without naming all of them, I should particularly mention Pat Easterling, Philip Hardie, Nick Jardine, David Konstan, Sachiko Kusukawa, Geoffrey Lloyd, Vivian Nutton, Christine Salazar, Malcolm Schofield, David Sedley, David Sider, Ineke Sluiter and Heinrich von Staden. A special note of thanks is due to Aude Doody, Laurence Totelin and Frances Willmoth for their patience, good humour and valuable suggestions as they each read the entire draft. Tamara Hug, Michael Coxhead, Seb Falk, Max Leventhal, Caroline Musgrove and Emma Perkins all helped me to improve the manuscript. At Cambridge University Press, all involved were always patient and helpful, particularly Kaye Tengco.

Newnham College was generous in support of my work, as was the Department of History and Philosophy of Science. My thanks are also due to the staff of the Classics Faculty Library, the Whipple Library and Topoi (Berlin), for their continuing helpfulness. Mary Jo Nye and Robert

Nye greatly encouraged me in this project, especially by inviting me to give the Horning Lectures at Oregon State University.

Various audiences, including those composed primarily of students, responded with insightful comments and questions to my work when it was presented in classes, colloquia and workshops. I am sad not to be able to thank Ian Mueller or Bob Sharples for their continued support; it was a privilege to have known both of them.

Earlier portions and/or versions of the following chapters have been published as noted here, and I am grateful to the editors, Andreas Kühne and Thomas Söderqvist, for making my contributions possible: Chapter 2, Taub 2008b; Chapter 5, Taub 2007.

Finally, Niall Caldwell contributed, as always, in ways too numerous to list here. I dedicate this book to him.

Spelling and Abbreviations

For the most part (but not always), I have adopted a 'latinised' spelling of Greek names and terms (e.g. Iamblichus rather than Iamblichos), to conform to general usage.

Anth. Pal.	*Anthologia Palatina*
ca.	*circa* = about, approximately
esp.	especially
FGE	D.L. Page (1981) *Further Greek Epigrams.* Cambridge.
fl.	*floruit* date; the time at which someone was assumed to have 'flourished', or been most active.
NH	Pliny the Elder, *Natural History*
KRS	G.S. Kirk, J.E. Raven, and M. Schofield (1983) *The Presocratic Philosophers.* Cambridge.
LSJ	H.G Liddell, Scott, R. and Jones, H.S. (1940) *A Greek-English Lexicon*, New (9th) Edition, with a Supplement, 1968. Oxford.
n.	note
OED	E.S.C. Weiner and J.A. Simpson (eds) (1991) *The Compact Oxford English Dictionary*, New Edition. Oxford.
s.v.	*sub verbo* = 'under the word' (to refer to dictionary entries)

Introduction

Today, if we wish to write or read about science or mathematics, we have many choices: in addition to traditional formats, such as journal articles and textbooks, we have the strikingly modern media of blogs and tweets. Each of these formats has particular formal characteristics: for example, contributors of research articles to the leading journal *Science* are limited to 4,500 words, while reports can only be 2,500 words long; the listing of 'keywords' is often required by journals.

Ancient Greek and Roman authors writing on scientific and mathematical subjects also had a range of choices available for conveying ideas and information. For example, writers on astronomy employed a number of different formats: Aratus of Soli composed poetry, Plutarch adopted the dialogue form, Claudius Ptolemy wrote extended, systematic prose works, Cleomedes circulated lectures, Geminus offered an introductory teaching text and Theon of Alexandria produced commentaries. All of these different textual formats were used to convey astronomical ideas and methods. This diversity of literary formats used for the presentation of information and ideas is, for modern readers, one of the surprising, yet seldom studied, features of ancient scientific writing. Each of these different formats entailed specific formal requirements for the author as well as expectations for readers.

Form matters, and it informs our understanding of the cultural contexts in which texts were produced and used. This book considers several important textual formats used by ancient Greek and Roman writers to convey scientific and mathematical information. This Introduction is meant to be an entrée to an approach to ancient scientific and technical texts.

By concentrating attention on the choice of medium used to convey the message, I consider the effects of literary conventions associated with particular literary forms on the presentation of scientific, mathematical and technical material. That scientific, mathematical and technical (including medical) information and ideas were communicated in a range of formats typically used for others purposes – including the literary – indicates that writing and reading about scientific matters was a part of broader Greco-Roman culture. Indeed, many of the texts considered here were not likely to have been written for specialists, but rather for a wider reading public, including students.

That said, it is important not to regard Greco-Roman antiquity as one homogeneous time or place: generalisations will always be of limited value. The authors considered here range in date from about 700 BCE to the sixth century CE. Their texts were composed in Greek or Latin, languages which had various dialects and changed during the course of antiquity. Furthermore, the various literary formats employed different registers and conventions, reflected in the language.

From looking at the different ways in which scientific ideas were communicated amongst the ancient Greeks and Romans, it is clear that the formats in which scientific work circulated were also not static: over time, new literary forms – such as the commentary and the encyclopaedia – emerged. At different points historically, certain forms dominated the scientific literature. Today, some ancient formats, including introductory texts and question-and-answer texts, continue to be used, while others, such as poetry, have been largely discarded in scientific contexts.[1] Intriguingly, some formats – such as the dialogue – were seldom used for scientific topics in antiquity but gained special prominence in the early modern period, with later authors looking back towards what they regarded as significant ancient models.[2]

By opting to write in a specific format, authors may have hoped to reach certain audiences: some types of texts are more appropriate to students, others to specialists. In addition to providing insights into how

---

[1] Erasmus Darwin's *The Botanic Garden* (1791), consisting of two poems (*The Economy of Vegetation* and *The Loves of the Plants*), is a modern example of a work of natural history presented as poetry; it was a publishing success.

[2] In the early modern period, two authors interested in promoting Copernican cosmology turned to the dialogue form. Galileo Galilei wrote two important works in the form of dialogues: *Concerning the Two Chief World Systems* (*Dialogo sopra i due massimi sistemi del mondo*, 1632) and *Two New Sciences* (*Discorsi e dimostrazioni matematiche intorno a due nuove scienze*, 1638). In 1608, Johannes Kepler produced a Latin translation of Plutarch's (born before 50 CE, died after 120 CE) *Dialogue on the Face on the Moon*, which he appended to his own work, the *Somnium* (*Dream*, published 1634); he explained that this latter work was, to some extent, an imitation of part of Plutarch's (Kepler (1870) *Opera Omnia* 8: 25).

specific authors sought to engage their audiences, the consideration of the choice of textual format also indicates how authors regarded their work in relation to that of others. By choosing to write a poem, an author may have sought to place himself within a particular tradition, for example, that of the archaic epic poetry of Homer and Hesiod, which brought with it great authority; in deciding to present ideas in a dialogue, authors may have intended to have their work read against the background of the philosophical milieu depicted in Plato's dialogues (Taub 2008a: 36–37; 76–77). Furthermore, the choice of format also reveals something about the community of authors and readers interested in scientific topics. Some authors (and readers) were specialist experts in particular fields; others were working from the standpoint of a less technical background, presumably appealing to readers with a general interest in their subject. So, for example, Aratus' astronomical poem the *Phaenomena* and the several Latin translations of that work composed in antiquity were not directed to specialist practitioners, but to a wider readership (Taub 2010). Given the range of options available to ancient authors, their choices reflected some authorial intentions, including a desire to reach particular audiences. Other aims might be to collect and share information, put forward a theory, argue against the ideas of others, gratify a patron, amuse friends or educate others, to name a few possibilities. The significance of such authorial choices in communicating scientific material in Greco-Roman antiquity is a central theme here.

Historiographical Context

Traditionally, historians interested in ancient Greek and Roman writings on nature and mathematics have tended to concentrate on the content conveyed in those works, relying primarily on the methods of intellectual history, often mining texts for the ideas contained there. This approach has borne substantial fruit, useful to classicists and ancient historians and also of great value to historians of philosophy, science and mathematics working in later periods; yet, it has its limitations, some of which have been hinted at earlier.

In recent years, historians have been increasingly concerned with issues relating to the presentation and reception of ideas, looking at rhetorical strategies and reading practices. Most of the scholarship to date has studied scientific texts from the early modern period and later (e.g. Blair 1999 on medieval and early-modern problem texts; Moss 1993 on sixteenth- and seventeenth-century texts; Rheinberger 2003 on the nineteenth and twentieth centuries). This research has demonstrated the desirability of looking beyond just the ideas contained in a written work to the modes of discourse adopted and the implications for understanding authorial intention and reception by readers.

Classicists have, for the most part, tended not to concentrate on those texts which convey scientific, mathematical and technical information and ideas, choosing instead to concentrate on more 'literary' texts. Technical texts – with few exceptions – are generally not regarded as 'literary' and – in the hierarchy of literary style – are often overlooked. In order to understand the place of these texts within the cultures in which they were produced and read, we need to think in terms of discourse, not just content.

The way in which technical texts are studied has begun to change. A few scholars working on ancient scientific texts have considered issues relating to literary form and authorial choice; much of their work to date has been published in German. In recent years, there has been a growing interest in studying ancient scientific and technical writings as *texts*, with scholars bringing to their study the same sorts of questions that classicists and literary scholars typically apply to more literary, non-scientific works.[3] Technical texts often become more accessible when approached in this way, being revealed as even more interesting and richly layered than they might seem at first reading. The study of scientific, mathematical and technical texts as texts (and not simply a concatenation of ideas) reveals crucial links to other aspects of Greco-Roman culture.

Literary Formats, Genres and Types of Texts

Before going further, it is important to clarify a few of the terms used here, as well as some of the working assumptions. The term 'genre' is an abstraction – a classificatory tool – used to describe different sorts of things, including films, paintings and forms of discourse (Paltridge 1997). The term is applied variously, as we might expect, by scholars in different disciplines.[4]

---

[3] See Doody, Föllinger and Taub 2012: 233–236 for a discussion of developments in the study of technical texts.
[4] Some historians refer to mathematics itself as a 'genre'; this usage of the term (to refer to an intellectual discipline) differs from the way it is typically used in literary and linguistic studies.

The medievalist Irma Taavitsainen, who works on medical linguistic corpora, has offered a useful definition of genres as 'inherently dynamic cultural schemata used to organise knowledge and experience through language'. Genres are based on conventions, and are realised in texts for specific purposes in certain cultural contexts. As Taavitsainen explains, genres 'become institutionalized so that they can function at the same time as "horizons of expectation" for readers', giving clues as to what to expect in a text, while providing models for authors. While genres are conventional, they are also fluid and dynamic, influenced by various factors culturally and historically. Indeed, some scholars regard genres as culture-specific. Genre is sometimes seen as being defined by 'external' cultural features, while another abstraction, the 'text type', is used to classify texts on the basis of 'internal' linguistic and structural features. Egon Werlich famously identified five 'types of texts': narrative, instructive, expository, argumentative and descriptive. But, as Taavitsainen points out, the terms 'genre' and 'text type' are often used vaguely – and even interchangeably – in the scholarly literature (Taavitsainen 2001: 139–140; Werlich 1976: 39–41). This is not a problem that can be solved here, but it is worth noting to highlight some of the difficulties – and lack of unanimity – involved in using these abstract categorisations.

Two classicists interested in questions relating to literary styles, Gian Biagio Conte and Glenn Most, use the term 'genre' to refer to 'a grouping of texts related within the system of literature by their sharing recognizably functionalized features of form and content'. They emphasise that genre is 'not only a descriptive grid devised by philological research, but also a system of literary projection inscribed within the texts, serving to communicate certain expectations to readers and to guide their understanding' (Conte and Most 2012). This view of genre points to choices made by authors (and compilers and editors) who have a number of options open to them as they craft their texts, and recognises also that readers have expectations that are shaped and served in different ways by different styles. Christopher Carey has noted that recent scholars have come to regard genres 'as flexible modes of communication with permeable boundaries', rather than as fixed (Carey 2007: 250; see also Most 2000: 17–18).

The aim here is not to try to un-pick the scholarship devoted to the study of such categorisations, but rather to adopt some of the tools – however problematic – used by literary and linguistic scholars, and others, to talk about texts. Nevertheless, it is important to note that the terminology employed by scholars is not applied uniformly; furthermore, terms like 'genre' are fluid and cannot always be easily pinned down. For our purposes here, David Duff has offered another good working definition of the term 'genre' as 'a recurring type or category of text, as defined by structural, thematic, and/or functional criteria'.[5]

Once we adopt the notion of genre, it may help to provide clues to the various aims of authors and to expectations of audiences of 'scientific' writings. For example, some scholars use the category of 'didactic poetry' to describe poems that were presumably intended to educate, or at least teach, while attracting the reader with the cultural values associated with metre; issues related to the use of the term 'didatic poetry' are discussed in Chapter 1. However, individual authors may have pursued a number of aims simultaneously: conveying specialist information, self-promotion, exerting intellectual and moral influence, entertaining readers and demonstrating allegiance to a particular point of view or social group are some examples (Doody, Föllinger and Taub 2012). In attempting to associate particular genres with specific authorial ambitions, we must guard against adopting some genre distinctions too enthusiastically: the reliance on a limited and particular group of categories (including 'didactic poetry' and 'treatise', to name two widely used terms applied to scientific and technical texts) suggests a degree of homogeneity which may result in the loss of finer-grained understanding of particular texts. A consideration of the formal features of texts often displays a greater and more interesting variety than we might otherwise expect, and in some cases reveals intriguing ambiguity. Some texts are genre-combining or -crossing, and may even be regarded as hybrids.

If we think further about terminology, it is important to note that text 'traditions' are also an abstraction in which one text builds on or is linked to another (Taavitsainen 2001: 150); this connection can be seen to be achieved in a number of ways, for example through subject matter or formal features, such as metre. At least one genre, that of commentary, involves deliberate crafting in relationship to earlier texts. The tradition of producing and using commentaries, begun in the Hellenistic period and often targeting literary as well as philosophical texts, served also as a conduit communicating earlier Greco-Roman ideas to other cultures, allowing later 'traditions' (for example, those of the medieval Jewish, Christian and Muslim scholarly communities) to engage with another, earlier one.

---

[5] Duff 2000: xiii; as he notes, the term is increasingly used to classify non-literary texts. Today, even Web pages are classified by genre; see Rosso 2008.

Defining 'Greco-Roman'

The term 'Greco-Roman' is intended to signal that intellectual work was undertaken across cultural and linguistic communities in the ancient Mediterranean world; many individuals, for example Marcus Tullius Cicero (106–43 BCE), had a working knowledge of both Greek and Latin. In many instances, activities pursued by those writing in Latin were closely related to work undertaken by Greek authors. This is not to say that some Romans, who had their own interests, agendas and approaches, did not on some occasions deliberately contrast what they regarded as 'Greek' ideas with Roman values. Nevertheless, Latin authors were often very self-conscious about their Greek predecessors and contemporaries.

'Science', 'Scientific', 'Scientist'

Literary and cultural terms – such as 'genre', 'tradition', 'Greco-Roman' – are not the only ones which present challenges: using the terms 'science' and 'scientific' to describe activities and texts in the ancient world is itself somewhat problematic. In part, the problem is due to the changing understanding – by modern practitioners, historians and philosophers – of what constitutes science. Indeed, the term 'scientist' is itself modern, said to have been coined by the polymath, historian and philosopher of science William Whewell (1794–1866) in 1840, in *The Philosophy of the Inductive Sciences founded upon their history* (Whewell 1840, vol. I, Introduction cxiii, cited in the *OED*); arguably, there is no exact analogue in other periods to that professional label. Historians, philosophers, sociologists and others who study the scientific enterprise have sometimes been much occupied by issues of definition, and although such issues cannot be addressed in detail here, they must be acknowledged (see Cunningham 1988). Nevertheless, there are numerous ancient authors and texts whose presence in any study of the history of science few, if any, historians would question.

What is labelled 'scientific' is often a matter of taste. There are some scholars, for example, who claim that the ideas of the Presocratic philosophers were not scientific, while there are numerous others who would counter that they represent the beginning of a 'proto-scientific' or 'scientific' enterprise; these positions cannot be argued here. Nevertheless, it must be acknowledged that the status of various human activities as scientific, non-scientific or even pseudo-scientific is not always clear-cut. While the ancient Greeks were themselves interested in classifying human activities, there was no one term, either in Greek or in Latin, equivalent to the meaning of the modern word 'scientific'. While some have suggested that the Greek word *epistēmē* is equivalent to the Latin *scientia*, this is debatable; furthermore, neither conveyed the same meanings as our modern word 'science'.

In the treatment of 'scientific' writings which follows, I have chosen to include works by authors – many of whom are usually considered to have been philosophers – who sought to explain nature (*physis*) and the physical world. I have also included mathematical writers, since many in antiquity who thought about such matters – including such diverse thinkers as Aristotle and Claudius Ptolemy – considered mathematics (*mathematikē*), like philosophy (*philosophia*), to be a branch of theoretical knowledge (*theoretikē epistēmē*). However, readers should be aware that ancient understandings of what constituted mathematics and philosophy do not map onto modern conceptions in a straightforward manner. Just as today there is not exact unanimity regarding what constitutes science and scientific practice, with some things being labelled as 'pseudo-science' or 'crank science', so in antiquity people were not unanimous about the categories in which they placed those pursuits which might today fall under the historian's rubric of 'scientific'.[6] As G.E.R. Lloyd has noted, 'a distinction between "philosophers" and "scientists" is in general hard to draw in Greco-Roman antiquity. Natural science is a domain that straddles both those disciplines as *we* perceive them'.[7] My working definition of 'science' is: 'an attempt to understand and explain physical phenomena'.[8] As with 'scientific', no one term, either in ancient Greek or in Latin, carried the same meaning of the modern English word 'science', with its primary reference to knowledge of the physical world. The Greek word *philosophia* means 'love of wisdom'; 'natural philosophy' can be understood as 'love of wisdom about nature (*physis*)'.

While amongst the ancient authors concerned with physical phenomena the meaning of *epistēmē* ('knowledge') could be variously understood and interpreted, Aristotle's view that *epistēmē* depended on knowledge of causes was very influential, not only in antiquity, but in later periods as well. According to Aristotle (*Metaphysics* 1025b25–6), *epistēmē* comprises

---

[6] On the topic of the classification of knowledge and expertise, see Tatarkiewicz 1963; Kühnert 1961.
[7] Lloyd 1991: 301, emphasis in the original. He goes on to distinguish (on p. 302) 'between those who engaged in detailed empirical work, and those who did not'.
[8] I use this working definition with the recognition that the meaning of 'natural' is, like that of 'science', not unproblematic and varies in different contexts.

three types of knowledge: the practical, productive and theoretical. Theoretical *epistēmē* can be further divided into three areas (*Metaph.* 1026a6–23): mathematics, physics and theology (metaphysics). Some types of *epistēmē*, as defined by Aristotle, have correspondences with what is understood as 'science' in later periods. With this in mind, I sometimes use the term 'natural philosophy' to refer to efforts to explain physical phenomena. Because ancient Greek philosophers used the word *physis* to refer to nature, and *physikoi* to refer to people who studied nature, I sometimes use the word 'physics' as an equivalent of the term 'natural philosophy'.

Describing Texts

There are many issues involved in the identification and description of different textual formats, types or genres. As has already been noted, these are large topics which cannot be dealt with fully here; nor can all the controversial questions be easily resolved. Nevertheless, it is important to recognise that the categories invoked are not entirely clear-cut and unproblematic. There are also issues encountered in identifying and describing texts as 'scientific' and/or 'mathematical', as it cannot be assumed that there is unanimity in understanding these descriptors as they are applied to modes of thinking, practice and texts. There is a danger of applying such terms ahistorically by suggesting that the modern usages map onto those of ancient authors and practitioners. Modern readers do not always agree as to what characterises a mathematical text; ancient authors who wrote about mathematics did not always agree on its definition either.

A certain style of presentation, with a systematic format, is often regarded by modern readers as the paradigm of ancient Greek scientific and mathematical texts. As Morris R. Cohen and I.E. Drabkin noted on the first page of their *Source Book in Greek Science*, 'the evidence indicates that the ideal of rigorously deductive proof, the method of developing a subject by a chain of theorems based on definitions, axioms, and postulates, and the constant striving for complete generality and abstraction are the specific contributions of the Greeks'.[9] This view of the characteristic style of Greek scientific explanation and mathematical thought has deep roots, going back to Aristotle in the *Posterior Analytics*, and to Plato before him. The influence of the 'axiomatic-deductive system' was powerful;

---

[9] Cohen and Drabkin 1948: 1. Of course, they are focusing on textual evidence; not much archaeological or material evidence survives.

significantly, Isaac Newton adopted an axiomatic-deductive style of presentation for his *Principia Mathematica*.[10]

The ideas and practices of ancient Greek scientific and mathematical work were presented in a wide variety of types of texts, for the most part in prose formats, but occasionally – as we will see in what follows – in poems. Some of these texts were written by *physikoi* and *mathematikoi*, men who presented themselves and were recognised by others as 'physicists' (or 'natural philosophers') and 'mathematicians'. However, many of the texts that we would identify as 'scientific' or 'mathematical' were written by individuals who were not themselves experts, some of whom might have been described (in a different time and place) as 'popularisers'.[11]

Even though there are numerous texts whose forms would today be regarded as appropriate and usual for scientific work, such as teaching texts and manuals, many ancient writings that are thought to be 'scientific' look very different from modern presentations of science and mathematics. Examples of Greco-Roman writing on scientific and mathematical topics are more rich and varied than we may have expected. This diversity demonstrates that a 'scientific' text is not always the same sort of thing, and identifying what characterises a 'scientific', mathematical or technical text may not be entirely straightforward. The formal diversity of these texts points to a more nuanced understanding of the place of scientific thinking in Greco-Roman antiquity, suggesting that it is not always simple to identify 'science', nor is it easy to pigeon-hole particular written works. Ultimately, broad generalisations may prove of very limited use. The aims of the author of a text and its intended function will often be best understood by investigating its specific historical context; this is an important undercurrent throughout this volume.

In the chapters that follow, several significant genres or types of scientific texts are treated in some detail. Here it seems appropriate to examine some examples of Greek terminology and their relationship to modern descriptions.

For example, many ancient scientific and technical texts are referred to as 'treatises' by modern scholars, yet it is not always clear exactly what this term means. In the modern period, prose is the dominant – even expected – form of scientific communication. Greek and Roman authors used prose as an important format for many scientific, medical and technical writings, even though it was not their only option. Some fifty years

---

[10] See Clarke 2011: 266, on Newton's *Principia* and axiomatic-deductive presentation.
[11] See, for example, Cuomo 2001: 73–79 on non-specialist authors writing about mathematics.

ago, Charles Kahn argued that the first Greek prose treatise was written in the sixth century BCE by Anaximander (Kahn 1960: 6–7 and 240). He regards Anaximander's *Peri physeōs* (*On nature*) as having been the first example of 'a new genre: the treatise *peri physeōs*' (Kahn 2003: 145; see also Graham 2010: 45, who agrees with Kahn). However, the question of what constitutes a 'treatise' remains to be addressed.

Prose texts had various formats; the dialogue was particularly favoured by Plato, but relatively few 'scientific' dialogues were composed in ancient Greece and Rome.[12] Other prose formats important in Greco-Roman antiquity are discussed in this volume, including the letter, commentary and encyclopaedia. Considering formal features, Philip van der Eijk has pointed to a 'less well defined species of text' that is 'sometimes referred to in modern terms as "treatise"'. The formal structure implied here does not fit into recognised ancient categories such as dialogue (*dialogos*), letter (*epistolē*), handbooks (*technai*), introductions (*eisagōgai*) or commentaries (referred to as *hupomnēmata*, by some ancient authors); as van der Eijk notes, the style of 'treatises' may be regarded as 'less elaborate' (van der Eijk 1997: 89). If we accept the lack of a distinct form which is attributable to 'treatises', it may make sense to read 'treatise' in these contexts as referring simply to a written 'work' rather than to a particular genre.

Kahn has posited that there was a tradition of technical prose writing, dating back to the invention of writing in Greece and the East, which included notes and memoranda used for the practice, improvement and teaching of technical work, such as astronomy, geometry, music, architecture and sculpture (Kahn 2003: 151); much of this writing would have been very specialised, probably incomprehensible to those not in the appropriate field. Acknowledging that there is not much evidence of such prose works, Kahn nevertheless advised readers 'to take into account the importance of what happens not to have been preserved', cautioning that the unknown and absent should not be mistaken for what never existed (Kahn 2003: 151).[13] Such notes may have been the first prose writings on technical topics in Greco-Roman antiquity, and would have included various sorts of *hupomnēmata* (here, to be understood as 'notes' or memoranda), including – perhaps – those that were aphoristic and memorable, useful to practitioners and specialists in a particular technical field.

---

[12] As was noted earlier, the dialogue was an important vehicle for presenting scientific explanations in the early modern period, as attested by the writings of Galileo Galilei.
[13] Nevertheless, in the absence of evidence, Kahn's suggestion remains speculative.

Portions of the medical writings found in the so-called Hippocratic corpus are amongst the earliest prose works in ancient Greece (Dean-Jones 2003: 112).[14] The works attributed to Hippocrates (most of which were not actually composed by him) show a capacity for accurate and concise statement, and are often 'aphoristic' (as, for example, recipes), that is, short, concise and often memorable statements. The first example in the book known as the *Aphorisms* in the Hippocratic Corpus is indicative of this type of text: 'Life is short, the Art long, opportunity fleeting, experiment treacherous, judgment difficult. The physician must be ready, not only to do his duty himself, but also to secure the co-operation of the patient, of the attendants and of externals'.[15]

Elizabeth Craik has drawn attention to the compilations of aphorisms within the Hippocratic corpus, noting that they are of the type designated by ancient commentators as *hupomnēmata* ('notes') in contrast to another type of text, the *sungramma* or *suntagma* (terms often translated as 'treatise') (Craik 2006: 335).[16] The term *hupomnēmata* can refer to notes that may serve as aides-mémoire.[17]

Kahn argued that a new form of technical literature, the 'Ionian prose treatise' – associated with the 'new' science undertaken by the Ionian philosophers – emerged from the assumed, older tradition of technical memoranda (Kahn 2003: 152). A new type of work, which we today sometimes refer to as the treatise, may have reached a wider audience than the *hupomnēmata* would have done, through the extended presentation of the topics covered.[18]

Given the power of poetry in the ancient Greco-Roman world, and the authority it conveyed, as exemplified by the canonical status of the Homeric and Hesiodic poems, the use of prose – that is, doing without metre and rhythm – begs for explanation. While it may have been assumed in the archaic period that metre was most appropriate for anything worthy of preservation, from the middle of the sixth century prose

---

[14] These texts, while regarded by modern scholars as being anonymously authored, were associated with Hippocrates in antiquity; this association may be regarded as a sort of 'branding' or 'badging'. See also Totelin 2004.

[15] [Hippocrates of Cos] *Aphorisms* 1.1, trans. Jones 1931: 98–99.

[16] Craik 2006: 342–344 also discusses the metre of some aphorisms, their mnemonic value, and relation to temple and to 'scientific' medicine.

[17] However, the use of the term *hupomnēmata* is not uniform across authors and periods.

[18] Furthermore, early 'treatises' – or more extended treatments – may have been read aloud. Kahn 2003: 152; see also Thomas 2003: 173–180. The oral presentation of written texts is a recurring feature of Greek and Roman culture; a number of scholars have suggested that Aristotle's 'treatises' began life as lectures; see Taub 2008a: 18.

literature began to be produced, breaking free from any assumption of the greater value and utility of poetic forms.[19] Indeed, the presentation of a text in a prose format which is meant to be shared with others has been understood by some as signalling a particularly significant cultural shift. The written preservation and transmission of some poems may have been secondary to their oral composition and performance; significantly, the archaic Homeric poems are understood to have been part of an oral 'tradition'.[20] In contrast, prose is normally regarded as being closely linked to literacy and to the culture of writing texts. Simon Goldhill has argued that in the fifth century BCE prose becomes 'the medium for authoritative expression, the expression of power' (Goldhill 2002: 5; cf. Kurke 2010: for example, 15, 47–48, on experiments and 'generic mixtures'). Of course, poetry did not suddenly disappear; on the contrary, it continued to wield great force. Yet, from a relatively early period, prose formats were important for the communication of scientific ideas and methods. And from the fifth century, prose was in the ascendancy as the primary mode of discourse conveying authority, particularly in certain fields.

Some of the earliest writers of Greek prose may have been living in Ionia during the sixth century BCE. Thales of Miletus is traditionally regarded as the first to have philosophised about nature, but it is not clear whether or not he wrote any works, or in what form they may have been presented. Even Diogenes Laertius, in his lengthy work on the *Lives of Eminent Philosophers*, expresses some scepticism regarding Thales' authorship of any work (1: 23). Anaximander of Miletus is traditionally regarded as a 'student' of Thales; only fragments of his work *Peri physeōs* (*On nature*) survive. As noted earlier, Kahn argued that Anaximander's writing is the earliest example of a new genre: the treatise *Peri physeōs* was characterised both by its subject matter (on 'nature') and the order of topics discussed, beginning with first principles and the origin of heaven and earth, and ending with a discussion of human beings (Kahn 1960: 6–7 and 240; Kahn 2003: 145).[21] But is the term 'treatise' helpful in understanding the character of Anaximander's writing?

---

[19] Thomas 1992: 64–65; also Andersen 1987. Thomas 1992: 57–61 depicted the diverse uses of writing in the eighth and seventh centuries, including examples of graffiti to mark ownership and protect objects.
[20] The scholarship on traditional oral poetry is vast. A good starting point is with the Parry-Lord thesis: see Parry 1971, Lord 1960, and Thomas 1992: 29–51.
[21] By this definition, the monologic section offered by Timaeus in Plato's dialogue of that name might qualify as a treatise *peri physeōs*.

The attraction of the term 'treatise' may be partly motivated by the fact that the prose treatise is an important form of modern scientific communication, and one with which modern readers are very familiar. The English word 'treatise' refers to a written work dealing formally and systematically with a subject.[22] However, 'treatise' is a modern term, and it is not entirely clear that there was an entirely equivalent term used by ancient Greek and Latin authors. Even a brief foray into the scholarly literature on scientific, medical and technical texts reveals references to a wide range of writings with different specific formats as 'treatises', suggesting that there is no characteristic form which defines the category 'treatise'; nor is the term applied in a homogeneous way. In actual usage, the word 'treatise' does not seem to relate to particular formal features, but often seems to be understood as a general term for a piece of writing dealing with a specified subject (such as *physis*).

Can the term 'treatise' be used to refer to a genre of text recognised in antiquity? While the term 'treatise' is often used to refer to a 'work' or 'text' in a general sense, the term also has a particular historical resonance and set of connotations in English, some of which are bound up with the history of philosophy and scientific discourse.[23] The Greek terms which are often translated as 'treatise' – such as *logos, pragmateia, sungramma, sungraphē, suntagma* – cannot be precisely translated by that term, because the English word 'treatise' conveys historically influenced meanings and expectations which do not neatly map onto the ancient Greek and Roman contexts in which the ancient works were written. Perhaps surprisingly to modern sensibilities – and further complicating the matter – the idea that a treatise need not be a prose work was voiced by Kahn, who described both poetical and prose examples of treatises *peri physeōs* (*On nature*), pointing to Parmenides, Empedocles and Xenophanes as having used verse as their medium (Kahn 2003: 145; Palmer 2009: 350 refers to Parmenides' poem as a 'treatise'). The character of such 'treatises' did not rely on the 'in metre' (*emmetros*) versus 'without metre' (*aneu metrou*) distinction, but was exemplified by the subject (*physis*) and the order of topics

---

[22] The first definition for 'treatise' in the *OED* is: 'A book or writing which treats of some particular subject; commonly (in mod. use always), one containing a formal or methodical discussion or exposition of the principles of the subject; formerly more widely used for a literary work in general'. Intriguingly, the online version notes that 'This entry has not yet been fully updated (first published 1914)'. Accessed 18 May 2015: www.oed.com/view/Entry/205390?rskey=9b3QH3&result=1#eid.

[23] David Hume (1711–1776), for instance, used both treatise and dialogue in titles of his own works, such as *A Treatise of Human Nature* (1739–1740) and *Dialogues concerning Natural Religion* (1779, published posthumously).

discussed. Kahn cites Hesiod's *Theogony* as sharing, in part, the same order as a treatise *peri physeōs*, crediting Anaximander with having produced a work which he describes as 'the prototype for a long-lived literary genre, the prose treatise *peri physeōs*' (Kahn 2003: 145–146). For Kahn, it is the content that defines the *peri physeōs* treatise, rather than formal features such as metre or prose.[24]

It may have been the case that ancient writers (and their readers) recognised a category of discourse concerned with a particular topic – namely *peri physeōs* – that included what are in our terms very disparate formats: prose and poetry. However, it is not clear that the modern English term 'treatise' can stretch to include poetry; certainly, the modern expectation of 'treatise' valorises prose formats. Although it may be important to note that different types of texts were recognisable as *peri physeōs*, it may not be helpful to label them all 'treatises', in our terms. The application of the term 'genre' to 'treatise' implies that treatises share some formal features, not only subject matter. Otherwise, we might well ask the question: Is Lucretius' *De rerum natura* a treatise?

A functional definition of a treatise *peri physeōs* as attempting an ordered account of a physical subject sets aside questions relating to formal and stylistic features of texts and embraces the possibility that such 'treatises' may have been presented in metrical or in dialogic formats, as well as in less elegantly crafted prose forms. While a possible genre of 'treatise' will not be discussed in further detail here, there are a number of Greek works, including those by Aristotle and Theophrastus, which can serve as examples of an ambition to provide a systematic discussion of a particular topic. The usefulness of referring to ancient types of texts (including poems and dialogues) concerned with the physical world as 'treatises' remains unclear, even while we may acknowledge that such works may have shared some features with other writings *peri physeōs*.

In the fifth and fourth centuries BCE, and later, a wide variety of prose texts were written on a broad range of subjects by various authors; many of these were rather technical. Franz Dirlmeier has suggested that the Greek word *pragmateia* refers to a written 'work'; his understanding of what constitutes a 'work' seems to coincide largely with common modern

---

[24] In the modern period there is an expectation that 'treatise' refers to a work in prose; cf., however, C.S. Lewis' self-claimed invention of the term 'treatise poem', cited in the *OED*, which also cites another example from the *Times Literary Supplement* in 1980 (www.oed.com/view/Entry/205390?rskey=o6o8p4&result=1&isAdvanced=false#eid; accessed 31 August 2015). Kahn 2003: 142 accepts the ancient view that Pherecydes' (*fl.* 544–541 BCE) work on theogony is the oldest Greek prose 'book'.

English usage of the term 'treatise'.[25] Dirlmeier was interested in issues related to oral and written discourse, and argued that the appearance of the systematic, thorough written work marked a development in styles of presenting philosophy *via* written texts. In his view, the term *pragmateia* signals work conveyed through a written text; he sees a shift from the dialogues of Plato, with their sense of oral exchange, to the *pragmateiai* of Aristotle.[26]

However, it is not clear that the term *pragmateia* refers to a particular text type,[27] format, or genre (if we understand that term to encompass form as well as content). Dirlmeier suggests that the term is a synonym for the written *logos*, indicating in particular serious, painstaking, almost professional mental activity,[28] but the word has other meanings as well, sometimes referring to a field of study or a practical endeavour.[29] In fact, the term *pragmateia* may refer to an activity as well as to a (written) product resulting from that activity. Dirlmeier suggested that the term 'work' (German *Arbeit*) reflects the meaning of *pragmateia* (Dirlmeier 1962: 10). In both English and German, work (*Arbeit*) can refer to an activity or occupation (even a profession), and also to the products – be they material or intellectual – of that activity (cf. Immerwahr 1960 on the use of the word *ergon* in Homer, Herodotus and Thucydides). We refer to the 'works' of an author, recognising that intellectual and other work was required for its production. We also refer to an author's 'treatment' of a subject; the word 'treatment' may indicate a finished product (such as a written 'work'), and also the process of treating the topic or question which is the subject of that treatment, or work. This duality of connotation, referring to both a product and the activity which results in that product, may be significant in understanding some uses of *pragmateia* (see Fischer 2013).

---

[25] Dirlmeier 1962: 10, where 'Arbeit' can refer to a 'written work'. Kahn 2003: 148 has suggested that *syngraphē* was the 'normal term for a prose treatise'; cf. Dover 1997: 183–184, who does not use the term 'treatise' as one of many possible definitions he offers of *sungraphē*. LSJ suggests that the term refers simply to what is written (for example, a narrative or a history).

[26] However, in this context it is worth noting that some scholars, including Föllinger 2012, see dialogical elements in Aristotle's extant writings.

[27] Using Werlich's term (1976: 39–41); see also Taatsavinen 2001: 140.

[28] Dirlmeier 1962: 10 'πραγματεία ein Synonym für den geschriebenen Logos ist und im besonderen die ernsthafte, mühevolle, ja geradezu berufliche geistige Betätigung anzeigt'. Lengen 2002 argues that Aristotelian *pragmateiai* do not all have same structure; see, for example, pp. 176–187 and 223–231.

[29] LSJ offers as definitions 'occupation' and 'business', citing Plato *Theaetetus* 161e referring to 'the business of dialectic' and Aristotle *Rhetoric* 1354b24 to 'the business of oratory'. See also Fischer 2013: 96–97.

Aristotle associates the term *pragmateia* with its purpose at the same time as referring to the text as a product (as at *Topics* 1.1 101a26, cf. 100a18).

While other terms, including *sungraphē*, *sungramma* and *suntagma*, refer to written texts, the word *pragmateia* is often used in scientific contexts (for example, referring to works on medicine, geography and music), and suggests a preoccupation with a specific branch of theory or practice. The term *pragmateia* does not seem to connote a particular style or format of writing, but rather a specialist and intellectualised approach concerned with a particular subject and related to specialised activities and practices, which may be conducted in a context of research and/or teaching (Fischer 2013: esp. 111–112; cf. Dirlmeier 1962: 9–11).

The term *pragmateia* can be applied to works in different literary formats, but formal characteristics are not the only features that define a genre. Indeed, as was noted earlier, 'genre' can be understood as 'a recurring type or category of text, as defined by structural, thematic and/or functional criteria' (Duff 2000: xiii; cf. Swales 1990: 45–58). The term *pragmateia* can be translated by the word 'work' to refer to the sort of written discourse that has the *function* to provide an account concerned with a particular subject, without the text adhering strictly to specific structural features. Significantly for us, some ancient *pragmateiai* dealt with topics understood by modern readers to be 'scientific'.

Chapter Outline

Each of the following chapters treats a particular genre, concentrating on a small number of target texts providing 'case studies'. Sub-themes which appear throughout the volume include the interplay between oral and literary culture reflected in many of the genres under consideration, and the ways in which particular genres offer information about intellectual communities in the Greco-Roman world, particularly those concerned with mathematics and explaining the physical world. For example, letters often give specific evidence of the relationships between the author and intended readers, including patrons, followers and members of correspondence networks; in particular, letters survive between members of the Greek mathematical community.

Chapter 1: Poetry

Poetry comported a special authority within the Greco-Roman world, even in texts devoted to scientific topics. Chapter 1 considers poems which explain physical phenomena or present mathematical problems. Scientific and technical writings were by no means confined to prose formats in Greco-Roman antiquity; in fact, a number of poems that reached very wide audiences in antiquity were on technical, scientific subjects. The oldest surviving Greek texts are epic poems; some of the earliest extant philosophical texts are also poetry.

The earliest ancient Greek writings – the Homeric and Hesiodic poems – were in metre, and, in the later Greco-Roman world, poetry was a particularly authoritative and respected format, adopted for a range of subjects and audiences. Poetry was an especially powerful format for ancient Greek and Latin philosophical and scientific texts.

Today, no scientist would choose to convey authority and expertise through publishing work in poetry: prose formats are the genres of choice for modern scientific communication. In sharp contrast to this, poetry must be included as an important form of scientific discourse in antiquity. Certain poems, for example Aratus' *Phaenomena* and Lucretius' *De rerum natura*, would surely be on everyone's top ten list of important poems written about the physical world in antiquity. Perhaps surprisingly, a number of poems presenting mathematical problems survive; I particularly consider these.

While the poems of Lucretius and Aratus are relatively well known, mathematical epigrams (for example, those preserved in the *Greek Anthology*) have been little studied and remain to be explained, from the mathematical as well as literary and more broadly cultural vantage points. Many of these poems display intriguing intertextualities with canonical authors (including Hesiod and Plato), and they require further attention in order for us to begin to understand their place in the history of Greek mathematics. Since these mathematical problem-poems preserved in the *Greek Anthology* are epigrams, we should consider the significance of that particular form. Significantly, 'poetry' was not a homogeneous category in antiquity. The ancient authors who wrote on literary texts differentiated between those composed in metre and those that were not; poetry was not a genre itself, but a category comprising different genres of poetry, including, for example, epic, elegy and epigram. As we will see, the distinctive features of epigram are particularly well-suited to the presentation of mathematical problems.

Chapter 2: Letter

Various sorts of letters were written and circulated in antiquity; not all letters were intended for private purposes. Some, like the letters published today in newspapers, were clearly meant for a wider readership. Some were not 'real' letters, but were attributed to fictitious authors, or spuriously credited to famous individuals, such as Plato. Letters were used for various purposes, including for giving philosophical advice and instruction; examples include the three letters of Epicurus (341–270 BCE) preserved much later in the 'Life of Epicurus' by Diogenes Laertius (probably first half of the third century CE). Other letters contained technical or scholarly work on mathematical, mechanical and medical topics. A number of letters written by ancient Greek mathematicians survive, indicating that letter writing was a useful mode of communication for them; for example, Eratosthenes of Cyrene (ca. 285–194 BCE), who worked in Alexandria, was the recipient of letters from Archimedes (ca. 287–212 or 211 BCE), living in Syracuse. Some of the letters are clearly communications between friends and colleagues, and have almost the flavour of a conversation; others, particularly the letters of Epicurus, were intended to be instructional, serving as brief summaries of his views for his students and followers. During the early modern period, dedicatory letters were often used to advertise patronage, but the practice dates to antiquity. In a letter attributed to Eratosthenes, directed to his royal patron Ptolemy III, the solution to a geometrical problem is announced. This particularly rich text is a focal point of this chapter.

Chapter 3: Encyclopaedia

Gaius Plinius Secundus (known to us as Pliny the Elder) was the author of a remarkable work, the thirty-seven-book *Historia Naturalis* or *Natural History* (='Enquiry into Nature'), which holds an important place in several fields of intellectual history, not least in history of science. The *Natural History* is usually referred to as an 'encyclopaedia'. It is the only one of Pliny's works that survives, and is a particularly interesting case, because, while other encyclopaedic writings may have been produced in antiquity (depending on how we define 'encyclopaedia'), the *Natural History* is the only one to have survived in its entirety. The *Natural History* stands at the beginning of a tradition that developed and flourished primarily in later periods. This chapter considers what it means to be an 'encyclopaedia', as well as looking at links between Pliny's work and Roman imperialism, concentrating on Pliny's treatment of 'scientific' material, arguing that his approach is both 'encyclopaedic' and imperial, whilst being characteristically 'Roman'.

Chapter 4: Commentary

As part of the developing literary culture of the 'book',[30] those working within didactic and scholarly traditions produced a variety of handbooks, epitomes and commentaries. Prose works and poetry (especially the Homeric poems) were both the subjects of commentaries; philosophical as well as mathematical texts were often the topic of such treatments. The works of Aristotle particularly attracted commentators, perhaps partly because of difficulties in understanding them (due in some degree to the nature of the corpus in which they were transmitted (see, e.g., Bodéüs 1993: 10–11).

While commentaries on various types of texts were important from the third century BCE, the commentary was a particularly significant genre for writing about scientific topics in the later period. The sixth century CE was an especially important period for the production and use of commentaries, and Alexandria was an important site for this tradition. The commentary continued to be a key genre for the medieval period, and flourished in Arabic, Hebrew and Latin.

Typically, a passage from the ancient source is quoted and then a comment appended, which may be of any length, from one sentence to several (modern) pages. And, the commentator may refer to other works, by the author of the target text or other writers. Even as commentaries encouraged a close engagement with particular texts, they often served as vehicles for the presentation of the commentator's own ideas. The commentary in some ways represents the culmination of the movement from oral forms of discourse to the establishment of new written traditions, which are in themselves text-focused. However, commentaries were not simply silent texts: they often were used within teaching contexts in which lectures and discussion took place; Porphyry, in his account of the life of his teacher Plotinus, reports that 'in the meetings of the school he used to have the commentaries read' (14.11–12; trans. Armstrong 1966: 41). A significant number of important ancient commentaries on scientific and mathematical works survive, including several on Aristotle's writings, as well as others on mathematical works, such as Proclus' commentary on Euclid. Three ancient commentaries on Aristotle's *Meteorology* survive, at least in part; these will be discussed in some detail here, as exemplars of the genre.

---

[30] See Knox 1989, Easterling 1989.

Chapter 5: Biography

Biography was not a clearly demarcated genre for ancient Greeks and Romans. The boundaries with other genres, including the eulogies or *encomia* used to praise heroes and important citizens, are not distinct and may be somewhat artificial. The *bios* ('life'; plural = *bioi*) as an account or celebration of a particular life can be found in a range of ancient writings and may also include a discussion of an individual's opinions (or *doxai*); such accounts often carried an ethical or religious message. While biography itself was not a rigidly defined genre for the Greeks and Romans, the *bioi* are linked by the desire to celebrate individuals.

Numerous *bioi* of ancient philosophers survive. One prominent subject is Pythagoras, who holds a special place in the history of science, since many later natural philosophers, mathematicians and scientists claimed descent from his intellectual line. In the twentieth century, Einstein argued that a scientist may even appear 'Platonist or Pythagorean insofar as he considers the viewpoint of logical simplicity as an indispensable and effective tool of his research'.[31] Three ancient accounts of the life of Pythagoras survive from late antiquity; they can all be loosely dated to about the third century CE, and were written by Diogenes Laertius, Porphyry and Iamblichus. Taken together, these 'lives' allow us to explore various strands of the ancient portrayals of the lives of individuals significant in histories of science, displaying formal resemblances to other important ancient texts, notably the Christian gospels.

---

[31] For the complete passage, see 'Reply to Criticisms', in Einstein 1949: 684; cited by Kahn 2001: 172.

I

Poetry

Let us begin our singing
From the Helikonian Muses
Who possess the great and holy mountain of Helikon...
> Hesiod *Theogony*, opening lines, trans. R. Lattimore

Blessed Pythagoras, Heliconian[1]
scion of the Muses,
answer my question:
How many in thy house are engaged in the contest for wisdom
performing excellently?
> *The Greek Anthology* Book 14.1, trans. W.R.Paton (5: 27)

If thou art able, O stranger, to find out all these things
and gather them together in your mind,
giving all the relations,
thou shalt depart crowned with glory
and knowing that thou hast been adjudged perfect
in this species of wisdom.
> [Archimedes] *The Cattle Problem*, trans. I. Thomas
> (1939–1941: vol. 2, 205)

Poetry was a powerful means of communicating in the ancient Greco-Roman world; poetry was often sung, as well as recited and read. The earliest surviving Greek works are poems, recognisable through certain formal structures, most notably metre. In ancient Greece, poetry was usually understood as being 'in metre' (*emmetros*), while discourse 'without metre' (*aneu metrou*) is prose (see, e.g., Plato *Phaedrus* 252b6; 258d10-11).

---

[1] *Heliconian*: of or relating to the mountain Helicon (or Helikon) supposed by ancient Greeks to be sacred to the Muses.

While different genres of poetry each have particular characteristics, resonances and functions, they share the fundamental feature of being composed in metre. The Athenian rhetorician Isocrates (436–338 BCE) noted that 'poets do everything in metres and rhythms, while prose-writers have no part in those means' (Isocrates ix. 10; trans. Dover 1997: 183).

Intriguingly, quotations of poetry within a prose text may function as a special kind of confirmation, or authoritative statement. The earliest Greek epic poets, Homer and Hesiod, are quoted and cited as authorities by later Greek and Latin authors, in all sorts of contexts and on all sorts of scientific and technical subjects for which we would not normally expect them to be considered experts. While the archaic poets may not have had the communication of scientific and technical information as their primary aim, portions of their work are incorporated as quotations or allusions by other authors to illustrate and support their own arguments.

Some readers particularly value poetry over prose formats. The Roman author Columella recognised this; in his *On agriculture*, he offered two books on horticulture: one in metre (Book 10), the other in prose (Book 11), each directed in response to a particular, named recipient who had stated a preference for prose, or poetry. In considering Greek and Roman poets writing about scientific and technical topics, perhaps the first question to be asked is: Why did these authors choose to communicate through poetry?

Of the numerous genres and forms of ancient poetry, several in particular were used to convey scientific and technical ideas and information, even mathematical problems. A number of ancient Greek and Latin poems have been treated by modern scholars as examples of 'didactic' poetry,[2] and the use of this term emphasises the presumed educational/instructional/pedagogical function of these works. Hesiod's *Works and Days* is often regarded as the first example of what might be called a 'didactic' poem.[3] However, it is not clear that ancient authors themselves recognised 'didactic' poetry as a distinct genre; as in epic poetry, hexameter was typically used in those

---

[2] There are debates about whether 'didactic poetry' is a genre itself; see Schiesaro 2012. Volk 2002: 34–43 refers to two late ancient authors who did consider it to be a separate genre: Diomedes (fourth/fifth century CE), author of an *Ars grammatica*, and the anonymous Peripatetic author of the *Tractatus Coislinianus*. For the purposes here, I cautiously follow Volk's lead and take a somewhat 'empirical' approach. Those poems that are primarily regarded as 'didactic' typically also display philosophical, political, religious, as well as literary concerns.

[3] See West 1966 and 1978; Blümer 2001: 107–260 argues that Hesiod may have been earlier than Homer, in terms of composition.

ancient poems now customarily read as 'didactic'. Poetry that is understood as 'didactic' presumably aims, among other things, to teach; there are notable examples of didactic poems devoted to various 'scientific' subjects, including those by Aratus and Lucretius. Yet while a desire to teach may have motivated the authors who composed certain poems, such as the anonymous *Aetna* and Manilius' *Astronomica*, in some cases there is clear evidence – from the poets themselves – that they had other motives too, including the entertainment of their readers and auditors.

In addition to 'didactic' poems, we also find epigrammatic poems that convey technical expertise, notably those in which mathematical problems are set out in a way that is simultaneously concise, erudite and playful, exactly in keeping with the features expected of this particular genre of poetry. Important examples of mathematical epigrams include the so-called *Cattle Problem* (*problema bovinum*) attributed to Archimedes, and the numerous mathematical problem-poems included in Book 14 of the *Greek Anthology* (*Anthologia Graeca*), a collection of epigrams ranging from the Classical to the Byzantine periods. Just as didactic poems aim to enlighten and often entertain, mathematical epigrams encourage intellectual engagement, offering challenging and clever riddles. Both didactic poems and epigrams are usually presented as free-standing texts, and they may also be quoted or cited by later prose authors. While these two genres of poetry may have somewhat related aims, they had very different formal features; so, for example, epigrams are brief, while many of the didactic poems are lengthy, sometimes composed over several books.

While scholars have devoted a good deal of attention to didactic poems on scientific subjects, in the modern period the mathematical epigrams are mainly commented on by historians of mathematics rather than classicists. Yet, the inclusion of those mathematical problem-poems that were brought together in late antiquity within the assemblage known as the *Greek Anthology* is evidence that they were valued as a particular type of intriguingly erudite and puzzling poem. And while the mathematical epigrams in the *Greek Anthology* are by no means homogeneous, considering them as a group as well as individually is helpful for understanding why they were valued both as poetry and as mathematics.

Furthermore, it is surely significant that collections of epigrams, problems and riddles were compiled in antiquity; these poems were not meant to be read and pondered only as individual and isolated texts.[4]

---

[4] On epigrams and anthologies see Cameron 1993, particularly chapter 1; elsewhere in the volume he provides the histories of specific manuscripts. See also Gow and Page 1965, vol. 1, Introduction; Gow and Page 1968: vol. 1; Page 1981.

Mathematical problem-poems were included in Book 14 of the *Greek Anthology* alongside other epigrams relating to oracles.[5] That the compiler of Book 14 included mathematical problems along with riddles and oracles requires us to consider the role of these mathematical problem-poems within wider Greek culture, a culture within which riddles were posed in various contexts (including different literary settings, such as tragedies) and had multivalent meanings.[6]

Intriguingly, a famous Greek riddle involved numbers, even if it is not actually mathematical: this is the riddle of the Sphinx, solved by Oedipus.[7] A short version is found in *The Library* (3.5.8), a collection of traditional myths, attributed to Apollodorus (probably second century CE). Athenaeus of Naucratis in Egypt (second century CE), in his *Banquet of the Learned* (οι *Learned Banqueters*=*Deipnosophists*), presents a longer version in Book 10, a book which is concerned with riddles more generally:

> Asclepiades in his *Stories Told in Tragedy* ... claims that the riddle of the Sphinx went as follows:
>
>> There is a creature upon the earth that has two feet
>>   and four, a single voice,
>> and three feet as well; of all that moves on land,
>> and through the air, and in the sea, it alone alters its
>>   nature.
>> But when it makes its way propped on the largest
>>   number of feet,
>> then the swiftness in its limbs is the weakest.[8]

A version of the Sphinx's Riddle, about the stages of a human life, also appears in Book 14 (as number 64) of the *Greek Anthology*, alongside other riddles and the mathematical problems.

Athenaeus devoted his work, the *Deipnosophistae*, to conveying the culture of the Greek symposium; he presents the riddle of the Sphinx as part of a larger discussion of riddles, their categorisation, and their place in

---

[5] Such as the oracle reported in number 65, allegedly given to Homer: 'There is an island, Ios, the fatherland of thy mother, which shall receive thee on thy death. But beware of the riddle of the young boys', as well as a number of riddles (*ainigmata/griphoi*). The riddle posed by the young boys is not included in Book 14, and is referred to in Book 7, 1.
[6] On riddles see Ohlert 1912; Schultz 1914; Gärtner 2008; West 2012. See Ceccarelli 2013: 248–249 on the place of riddles and oracular pronouncements – traditionally composed in hexameter – in plays, including the so-called 'riddle comedies'; Antiphanes' *Problema* may have been such a comedy, in which the solving of riddles played a central role.
[7] Intriguingly, the riddle is mentioned, but not itself posed, in Sophocles' *Oedipus the King*.
[8] Athenaeus *Deipnosophistae* 10.456b-c (trans. S.D. Olson); cf. *Greek Anthology* vol. 5, 14.64 and Apollodorus *Library* Book 3. See also Rokem 1996.

the culture of symposia, in which the performance of poetry – including riddling poems – was an important feature. As a form of discourse, the riddle has a kinship with oracular questions, as well as with intellectual problems of the sort posed in epigrams during symposia. Book 10 of the *Deipnosophistae* depicts the riddles presented at the symposium, and can itself be read as a collection of riddles. Athenaeus appears to have used as a source Clearchus of Soli's (fourth century BCE) *On Riddles* (*Peri griphōn*), of which only fragments now survive, but which was an earlier collection of examples.[9] Indeed, Athenaeus' work and the *Greek Anthology* together comprise the principal collections of Greek riddles now extant; as already noted, the mathematical epigrams are juxtaposed within the same book as the riddles and oracles, in the collection of the *Greek Anthology*. Like the mathematical epigrams, riddles were also typically presented in verse, including hexameter, the oldest known Greek metre, that of epic poetry, of the Homeric and Hesiodic poems.

While not a riddle per se, an ancient story of a question-and-answer game or contest featuring numbers and counting is depicted as taking place between Homer and Hesiod. Among the questions 'Hesiod' poses is the following: How many Achaeans went to Troy together with the sons of Atreus?. 'Homer' answered with a calculating-problem (*logistikon problema*): there were fifty hearths, and in each there were fifty spits, and on each fifty pieces of meat, and three (times) three hundred Achaeans were around each piece of meat. Sadly, the original text does not survive, but there is an attempt – possibly by a later annotator – at a calculation. Intriguingly, this story about who is the best of the two famous poets includes a mathematical problem, giving a strong signal that mathematical poems were embedded within broader Greek culture.[10]

With this in mind, before turning to the mathematical epigrams in more detail, a brief consideration of the use of poetry to communicate philosophical, scientific and technical ideas in the Greco-Roman world is in order, to better understand the background against which these texts were composed, read and recited.

[9] Clearchus' collection of riddles may have been in part inspired by other collecting practices undertaken within Aristotle's school; see Kwapisz 2013.
[10] *The Contest of Homer and Hesiod*, in West 2003: 333–335; Allen 1912: vol. 5, 225–238, esp. 231, lines 140–150. The dating of this compilatory work is not straightforward (West 2003: 298–299); the present form incorporates material from the fourth (and possibly the fifth) century BCE. West suggests that the calculation is due to a Byzantine annotator; he offers a figure of 112.5 million for the number of men. Naiden 2013: 272 points to the jokey character of response, noting that it provides no useful information about meat production or consumption. I thank Jochen Althoff for reminding me of this arithmetical problem. See also epigram 147 in *The Greek Anthology* (in Appendix A).

The Early Epic Poems: Traditional and Authoritative

The earliest extant Greek texts – the Homeric and Hesiodic poems – are epic poems which provide formal templates as well as backdrops for much of what follows, both in Greek and Latin literary culture, regardless of the topic. Epic metre (dactylic hexameter) was the recognisable shape of the discourse of these most ancient authorities, the epic poets, and it became the most important metre used by classical Greek poets, as well by many of the later ancient Greek and Latin poets – including Lucretius and Virgil – who offered explanations of the physical world. In using this metre, these poets associated themselves with the epic tradition.[11]

But metre was not the only legacy of the epic poets. There was a tradition of regarding the earliest poets, Homer and Hesiod, as great thinkers. Standing at the fountainheads of tradition, they helped shape intellectual agendas.[12] Several of those usually regarded as the earliest philosophers (lovers of wisdom) expressed their ideas in epic metre, Parmenides and Empedocles being particularly important examples. Understandably, modern scholars interested in these philosophers have tended to focus on their ideas, and almost seem to have taken for granted their use of poetry as the medium of communication, without commenting on their use of the metre of epic (a notable exception is Mourelatos 1970/2008).[13]

In the opening to his book *Nature in Greek Poetry*, George Soutar averred that 'Homer – if one may use the name as a convenience, without discussion, and without necessarily assuming identity of authorship for *Iliad* and *Odyssey* – is the mouthpiece of an age and a race. He throws the glamour of poetry over an early stage of Hellenic life, the life heroic' (Soutar 1939: 1). This passage from Soutar reminds us that when we use the names 'Homer' and 'Hesiod' we are using a form of shorthand to refer to 'authors' about whom we know very little. Since antiquity, the dating of the poems ascribed to Homer and Hesiod has been a topic of discussion; many modern scholars date the Homeric poems to the eighth century BCE, judging the Hesiodic poems to be somewhat later.[14]

---

[11] For an excellent introduction to metre, see West 1982.
[12] Plato (*Cratylus* 402b; *Theaetetus* 152c, 180c-d) and Aristotle (*Metaphysics* 983b27-32) both suggested, perhaps somewhat jocularly, that Hesiod and Homer were the fathers of philosophy. On ancient readings of Homer, see, as a start, the contributions in Lamberton and Keaney 1992.
[13] See, for example, Palmer 2009, who repeatedly notes that Parmenides wrote a poem, but has little to say about Parmenides' use of poetry. Hexameter is both the metre of epic and the metre of 'didactic' poetry.
[14] But Blümer 2001: 107–260 argues that Hesiod's work was earlier.

Hesiod's *Theogony* is concerned with the physical world and its generation, and his *Works and Days* briefly touches on technical matters. While the Homeric poems are not concerned primarily with the physical world or technical topics, they nevertheless conveyed information which was taken very seriously, over a very long period of time. And the Hesiodic and Homeric poems certainly carry a sort of glamorous authority, which is clear in the many quotations and allusions to them, even in prose works on technical subjects.[15] For example, in his prose work the *Meteorology*, in his discussion of the changing relationship between water and land, Aristotle mentions Homer as something of an authority on the geography of Egypt (*Meteorology* 351b32-352a2). Reverence for the archaic poets was not restricted to the Greeks, or to Homer; Pliny the Elder (died 79 CE) considered Hesiod (whom he regarded as the Father of Agriculture) to be as valid an authority on celestial matters as specialist astronomers (Pliny *Natural History* 18.201; cf. 18.212-3).

Not only were the ideas and information in the earliest epic poems held in great regard, but the form of these poems – epic hexameter – as well as other stylistic features provided models for later authors, no matter what their subject. Poetry was a key form of communication in the Greek and Roman worlds, and composing and reciting poetry was an esteemed activity; the Greek verb *poiēsis* literally means 'making'. Furthermore, experiencing the poetry of others – whether as an auditor or reader – was also highly valued. Poetry could be enlightening; it could also be entertaining (Dover 1997; Goldhill 2002).

What Soutar referred to as the 'glamour' of epic poetry, as well as its authority, ensured that metre was a powerful medium of communication for all sorts of ideas. This is evident, for example, in the poem composed by one of the earliest philosophers, Parmenides of Elea (born *c.* 515 BCE), who presented his ideas in a single and extraordinarily influential work, a poem which survives only in fragmentary form.[16] The title under which it has been transmitted, *On Nature*, may not be authentic, and our knowledge of the contents largely comes to us through the testimony of other ancient authors rather than through surviving fragments of the poem. According to the ancient reports, Parmenides attempted to explain a very

---

[15] Aude Doody has raised the question (personal communication) whether the use of poetry in quotations within a wider prose work is a different gesture from writing the whole thing in verse.

[16] Preserved largely by Sextus Empiricus (second century CE) and Simplicius (sixth century CE); see KRS 1983: 241. The Greek colony of Elea was located in what is now southern Italy. On Aristophanes' *Thesmophoriazusae* as a philosophical comedy drawing upon Parmenides' ideas, see now Clements 2014.

wide range of natural phenomena, including the origins and behaviour of the heavenly bodies as well as things on earth (see, for example, KRS 1983: 255–262). Alexander P.D. Mourelatos began *The Route of Parmenides* by pointing out that 'What we have from Parmenides is not a philosophical treatise but a poem in hexameters. On the face of it, the very form of the work places him in the tradition of Greek epic poetry' (1970/2008: 1). M.R. Wright also emphasised the use of epic metre, style and language by Parmenides (and Empedocles), suggesting that this was a deliberate authorial choice to 'ease the reader or listener into an unfamiliar and complex message by way of the old formulae', setting 'the traditional language of Homer and Hesiod in a new context' in which 'literary devices were called to mind but then deliberately subverted' in the exposition of the new philosophy of nature (Wright 1998: 22). Catherine Osborne has countered Wright's reading, arguing that we must understand Parmenides and Empedocles as poets who, like the earlier epic poets, 'formulate their thought directly in the familiar language of poetry'. Poetry is not the '*medium* for something else, their philosophy'; in her view, poetry was the 'default setting for these thinkers' (Osborne 1998: 25–26, emphasis in the original).

From Prose to Poetry?

However, in some other (later) cases, it might be argued that poets 'translated' ideas from the prose works of others into a sweeter, more palatable format; for some, poetry was the deliberate choice for communicating what might be regarded as difficult information and ideas. For example, Aratus' *Phaenomena* and Lucretius' *De rerum natura* (*On the nature of things*) both have close links to earlier prose works, by Eudoxus and Epicurus, respectively. Aratus is understood to have taken Eudoxus' prose work as his source, which he 'set' to verse. Lucretius elaborates on Epicurus' ideas and arguments, but does not follow him slavishly; David Sedley has argued that Lucretius was aware that he could not simply 'translate' Epicurus' ideas into Latin poetry (Sedley 1998). Historically, Aratus' and Lucretius' poems are amongst the most widely read poems written about the physical world, but their prose antecedents are barely known. Eudoxus' *Phaenomena* is encountered today only through Aratus' poem and a commentary by Hipparchus; Epicurus' *On nature* is lost, and his extant writings (mainly letters) are not as widely read as Lucretius' poem. There is a question as to whether the popularity of these works was determined by their status as poems, with poetry offering a vehicle for fame.

'Didactic' Poetry

These poems and others, including Manilius' *Astronomica*, are sometimes treated primarily as 'didactic' poetry by (some) ancient as well as modern authors, and share specific characteristics, most notably an explicit intent to teach.[17] Some ancient authors voiced their opinions that learning might be more achievable if presented in an appealing – even a fun – manner. Twice in *De rerum natura*, Lucretius stated that he thought using poetry as a medium of communication might make his subject – Epicurean philosophy – more palatable to his readers. He drew an analogy to the way physicians coat the rims of cups with honey to persuade children to take medicine (1. 921–950 and 4.1–25, in almost the same words). Other important intellectuals, notably Plato, also express the view that learning should be fun and enjoyable (*Laws* Book 7, 819). Questions about any presumed didactic function and entertainment value are particularly relevant to the mathematical poems, and are worth keeping in mind as we look at them more closely.

Epigrammatic Poetry

The original sense of the term 'epigram' indicates that the composition was intended to be engraved or inscribed. The earliest epigrams were inscribed on funerary monuments and votive offerings, used in Greece from an early period – not only as the vehicle of personal feeling but as the recognised commemoration of remarkable individuals or events. For this reason, epigrams are sometimes regarded as a type of occasional poetry, marking something particularly noteworthy.

When inscribed on stone or metal, such compositions are normally necessarily brief, not least because of the material and labour involved. Epigrams survived the shift from stone to papyrus, and their subjects were broadened, becoming more diverse. Epigrams were also often performed or recited, for example, at symposia, satisfying a desire for entertainment as well as intellectual stimulation. Indeed, Niall Livingstone and Gideon Nisbet argue that the epigram is the defining literary genre of the symposium (Livingstone and Nisbet 2010: 4; Bruss 2010: 124–129).[18]

---

[17] Palmer 2009: 318 has referred to Parmenides' poem as 'didactic', but this may put undue emphasis on the educational intent of Parmenides.

[18] See Bing and Bruss 2007 for bibliography. Cameron 1995: 71–103 argues that non-inscriptional epigrams from the Hellenistic period onwards were linked to performance at symposia.

Nevertheless, as the literary expression of the epigram detached itself in form from its inscriptional origin, certain stylistic restraints endured: conciseness remained key, and became a hallmark of the genre. From the standpoint of form, the epigram may itself have encouraged particularly active sorts of reading. Peter Bing has suggested plausibly that the reading of inscriptional epigrams involved the active taking in of the physical context of the inscription, its 'setting'. Epigrams read on papyrus (or heard at symposia) also encouraged (sometimes required) activity on the part of the reader or auditor. Bing opines that 'in concert with the genre's traditional concision, the extent of the reader's role in constructing meaning exceeds what is found in other genres' (Bing 1998: 38).

Akira Yatsuhashi has pointed to several salient features of the epigrammatic genre. The literary epigram was developed during the fourth century BCE, and the genre came to serve various functions. Yatsuhashi explains that the 'literary epigram became a primary means of disseminating and producing cultural knowledge and tradition'. The brevity and allusive character of epigrams 'made nearly every word in each epigram vital towards deciphering them'. Characteristically, the epigram 'demanded that its audience have control over a large body of poetical and cultural minutiae to comprehend the references contained in each. Epigram rewarded those who had acquired that base of knowledge or who had access to it' (Yatsuhashi 2010: 150). Those mathematical poems we have from ancient Greece are presented in the concise format of the epigram, and depend on the involvement of the reader to be fully understood: the reader provides a solution to the problem posed.

Of course, ancient audiences for epigrams would have been familiar with other genres of poetry; there are, in fact, many allusions in epigram to other poetic forms, particularly heroic epic (Bing and Bruss 2007; Gutzwiller 1998; Harder, Regtuit and Wakker 2002). In some cases, these allusions can be understood as grounding the poems in the familiar, almost nostalgically, with references to ideas and images already known to the reader or auditor. As the epigram developed as a poetic genre in the Hellenistic period, many ended with a 'twist' or a 'point'; this type of ending made the epigram particularly suitable for riddles, oracles and mathematical problems, offering a question or problem to the reader or auditor, which must be solved for the epigram to be fully understood. In this way, mathematical epigrams can be contrasted with the 'didactic' poetry of, for example, Aratus and Lucretius. Rather than aiming to provide a description or explanation of the physical world, the mathematical epigrams pose problems to be solved. And typically, no solution was offered, leaving the reader to puzzle it through. In order for the reader or auditor fully to comprehend the mathematical epigrams, they must engage in mathematics, they must 'do' maths.[19] (Whether or not readers would have been expected to consult other texts, or people – including teachers – to arrive at a solution is not clear.)

Mathematics as Poetry

That mathematics was a subject of Greek poetry may be especially surprising because a particular style of text, in a systematic format, is often seen as the hallmark of Greek mathematics. Indeed, as has already been noted, on the very first page of *A Source Book in Greek Science* Cohen and Drabkin suggest that the ideal mathematical text is the rigorously deductive proof, exemplifying 'the method of developing a subject by a chain of theorems based on definitions, axioms, and postulates, and the constant striving for complete generality and abstraction' (Cohen and Drabkin 1948: 1). Yet, upon further examination we see that the ideas and practices of ancient Greek mathematics were presented in a wide variety of types of texts, for the most part in prose formats, but occasionally in poems (see Taub 2013). Certainly, poetry can be seen to be as formally 'systematic' as mathematical prose, for it too is distinguished by set conventions and formats, for example those relating to metre.

Some of those texts we identify as 'mathematical' were written by *mathematikoi*, men who presented themselves and were recognised by others as 'mathematicians', including the renowned Archimedes. The *Cattle Problem* attributed to Archimedes (*c.* 287–212 or 211 BCE) is noteworthy for the difficulty of the mathematics involved; a seemingly simple and practical problem about counting herds of different-coloured cattle is set out in metre. The problem requires that eight unknown quantities be found. It was satisfactorily solved only in the twentieth century, and the solution required the use of computers; it is not known whether the ancient author of the poem thought it was soluble and, if so, by what means. A question for us is: Why was the problem presented in verse? (Intriguingly, we know of no setting out of the problem in prose in antiquity, a format that we would expect to be more typical for the presentation of mathematical problems.) We will return to the *Cattle Problem* later.

---

[19] This is in contrast to the experience of reading some mathematical texts, for example the *Elements*, which may permit, on some level, a more passive reading.

A number of mathematical problem-poems can be found in the *Greek Anthology*. The *Greek Anthology* contains more than forty mathematical problems presented as epigrams[20]; many of these are thought to have been collected by Metrodorus of Tralles, who is usually referred to as a grammarian (*c.* 550–600 CE, though his dates are not certain; Albiani 2006: vol. 8, 839 (9)).[21] Historians have conjectured that the mathematical problems would have been devised (and even the poems possibly composed) much earlier (Cohen and Drabkin 1948: 25–26, n. 1). This raises the question of why a grammarian would be interested in mathematical material.

Some of the texts that we would identify as 'mathematical' were written by people who were probably not specialists in mathematics (see Cuomo 2001: 73–79) The mathematical riddles presented in metre in the *Greek Anthology* are examples. In what follows we examine individual examples of mathematical epigrams in some limited detail, with a view to understanding why these problems were presented in that form, who their intended audiences may have been, and what these poems were intended to convey and accomplish.

But not all of the mathematical problems-as-poems from ancient Greece require the mind of Archimedes (or a computer) to be solved. Many of the mathematical problems in the *Greek Anthology* appear at first reading to be simple and mundane, requiring the counting of pieces of fruit or numbers of bowls of different metals, yet several are mathematically sophisticated. And even a brief consideration of these mathematical epigrams shows that many had correspondences with well-known myths and legends; several have clear intertextual references to other poetry, including the Homeric and Hesiodic poems, as well as allusions to myth more generally. Indeed, it could be argued that, given the abundant cultural allusions and references to myths, formal intertextuality is not really necessary, particularly when references to myth are readily accessible in vase paintings and other art forms. The allusions to myth may not be formally intertextual but may nevertheless reflect the pervasiveness of particular cultural features and values.[22]

---

[20] Forster 1945: 43 counts forty-four, as do Cohen and Drabkin 1948: 25 n. 1, whereas I count forty-six in Book 14, as does Heath 1964: 113. Clearly, we do not all agree as to how to count these problems or on what is to be counted.

[21] While not much is known about mathematical education in antiquity, basic mathematical skills may have been taught by those who were also teaching literacy. See Marrou 1956: 150–158; Sidoli 2015; Cribiore 2001: 180–183 on education. The word 'mathematics' can be understood as 'things to be learned'; the Greek noun *mathēma* derives from the verb *manthanō*, 'I learn'.

[22] There is a degree of fluidity in myth; several versions often coexist. Strikingly, this is also true for some riddles (as with that of the Sphinx, mentioned earlier). Furthermore, in some cases (again, including the Sphinx) the textual versions of particular myths, as well as riddles, are relatively late.

In fact, there are other epigrams in the *Anthology* which, though not being mathematical problems, encourage readers to count. Leonides (sometimes spelled Leonidas) of Alexandria, allegedly a former astrologer or astronomer during the reign of Nero (54–68 CE), composed a number of isopsephic (equal pebbles, used for counting) poems. More than forty of these survive, several having been offered as birthday gifts to members of the emperor's family, indicating patronage links. In ancient Greece, the letters of the alphabet also had numerical value; every unit (1, 2, …, 9) was assigned a letter (alpha=1, beta=2, and so on). Each decade was also assigned a letter (10=iota, and so on), as was each hundred (with rho=100 and the last letter of the alphabet, omega=800).[23] Every letter in a word, or line of verse, had a numerical value; Leonides' isopsephic poems allowed the reader to add up the numerical values of the letters to find that subsequent verses were equal in value. He explains (*Anth. Pal.* 9.356=*FGE* Leonides 33): 'We broach the drink of a different fountain in order to enjoy the marvelous poetry of the bard Leonides. For the couplets equal each other in their tallies. But you, Blame, begone, and harm others with your sharp bite'.[24]

Leonides' epigrams invite readers actively to convert the words and letters of each verse into numbers, which are then counted and compared to those of another. The numbers that result are typically large (in the thousands), and it is easy to imagine that the challenge of counting so high also offered some special pleasure; that large numbers were fascinating is also attested by the work of mathematicians, including Archimedes in his *Sand-Reckoner*. While Leonides' epigrams do not require a great deal of mathematical skill or experience, they do rely on the reader's active engagement, and presume a degree of knowledge about numbers and counting. And, as is typical for this poetic genre, there is a degree of intellectual challenge and playful surprise.

One of the texts considered here is alleged to have been written by a celebrated thinker, Archimedes, but his authorship is not certain. There has been the suggestion that it is an ancient 'forgery', a text written in antiquity and ascribed to a famous person, as a way to attract attention and possibly to gain the authority and lustre of the alleged author. Here, there are parallels to other ancient texts to which the name of a famous person has been attached. For example, Laurence Totelin has explained

---

[23] The twenty-four-letter alphabet was extended by using three obsolete letters: *digamma* ϝ (or, later, ς) for 6, *qoppa* ϙ for 90, and *sampi* ϡ for 900.

[24] Trans. Livingstone and Nisbet 2010: 119; see also Page 1981: 503–541, on Leonides.

that, in the case of the pharmaceutical known as 'Mithradates' Remedy', certain Romans used the royal name to create a 'Roman' drug, the recipe for which may have had no real association with King Mithradates VI of Pontus at all (Totelin 2004; see also Mayor 2010). This is arguably a similar case of associating a solution to a problem (medical or mathematical) with the name of a famous person to gain credibility and esteem. Some scholars have questioned whether Archimedes was the author of the poem, but most assume that he was familiar with the problem presented therein.[25] Here I generally refer to the composer of the *Cattle Problem* as Archimedes, recognising that it may also have been written and presented by someone unknown to us.

The *Cattle Problem*

Archimedes, a mathematician of the greatest renown in antiquity, is credited with the poem which presents the mathematical problem known as the *Cattle Problem*. In this poem a problem is posed, almost as a riddle, but no solution is offered.[26] The text opens in the following way: 'A Problem [*problēma*] which Archimedes devised in epigrams [*epigrammata*], and which he communicated to students of such matters at Alexandria in a letter to Eratosthenes of Cyrene'.[27] We know from other sources that Archimedes, living in Syracuse, corresponded with Eratosthenes, the Librarian in Alexandria; he apparently addressed at least one other work, the *Method concerning Mechanical Theorems*, to Eratosthenes.[28]

The *Cattle Problem* is presented as a poem, forty-four lines (in elegiac couplets) in the modern edition. The mathematical problem is itself tightly woven into the poetic format. As was noted earlier, no ancient prose setting of the mathematical contents of the *Cattle Problem* is known. Here is the opening of the English prose translation by Ivor Thomas (1939–1941: vol. 2, 203–205):

> If thou art diligent and wise, O stranger, compute the number of cattle of the Sun, who once upon a time grazed on the fields of the Thrinacian isle of Sicily, divided into four herds of different colours, one milk white, another a glossy black, the third yellow and the last dappled.

---

[25] See Heath 1921/1981: vol. 2, 23. Netz 2009a: 146, 168 assumed that Archimedes was responsible for the poem itself.

[26] The text of the *Cattle Problem* was discovered and edited in 1773 by Gotthold Ephraim Lessing, librarian at the Herzog August Library in Wolfenbüttel.

[27] Trans. Thomas 1939–1941: vol. 2, 202–203, with slight emendation; in what follows here, the prose translation of the poem is from Thomas 1939–1941: vol. 2, 203–205.

[28] Heath 1912: 5–51, 'The Method of Archimedes' (published as a supplement to his 1897 volume, with the pagination beginning anew, following p. 326). See also Chapter 2 here, on letters.

The poet goes on to explain that

> In each herd were bulls, mighty in number according to these proportions: Understand, stranger, that the white bulls were equal to a half and a third of the black together with the whole of the yellow, while the black were equal to the fourth part of the dappled and a fifth, together with, once more, the whole of the yellow.

The recipient is told to

> Observe further that the remaining bulls, the dappled, were equal to a sixth part of the white and a seventh, together with all the yellow.

The proportions of the herds are then presented as follows:

> The white were precisely equal to the third part and a fourth of the whole herd of the black; while the black were equal to the fourth part once more of the dappled and with it a fifth part, when all, including the bulls, went to pasture together. Now the dappled in four parts were equal in number to a fifth part and a sixth of the yellow herd. Finally the yellow were in number equal to a sixth part and a seventh of the white herd.

In other words, the recipient of the problem is invited to find the number of cattle (male bulls and female cows) in each coloured herd, that is, to find eight unknown whole numbers (there being no fractional cattle).[29] Following the listing of the relevant proportions for each herd, the reader is then told:

> If thou canst accurately tell, O stranger, the number of cattle of the Sun, giving separately the number of well-fed bulls and again the number of females according to each colour, thou wouldst not be called unskilled or ignorant of numbers, but not yet shalt thou be numbered among the wise.

Having solved the first part of the *Cattle Problem*, another, rather different type of solution remains to be found. The first part of the problem states

---

[29] Heath 1912: 319 (followed by Thomas 1939–1941: vol. 2, 205–206) provided modern equations to illustrate the proportions, summarized here:
Let W, w be the numbers of white bulls and cows respectively,
X, x be the numbers of black bulls and cows respectively,
Y, y be the numbers of yellow bulls and cows respectively,
Z, z be the numbers of dappled bulls and cows respectively.
$W = (1/2 + 1/3) X + Y$
$X = (1/4 + 1/5) Z + Y$
$Z = (1/6 + 1/7) W + Y$
$w = (1/3 + 1/4) (X + x)$
$x = (1/4 + 1/5) (Z + z)$
$z = (1/5 + 1/6) (Y + y)$
$y = (1/6 + 1/7) (W + w)$.

proportions relating eight unknown whole numbers; the second part goes on to describe the two groupings of the bulls:

> But come, understand also all these conditions regarding the cows of the Sun. When the white bulls mingled their number with the black, they stood firm, equal in depth and breadth, and the plains of Thrinacia, stretching far in all ways, were filled with their multitude. Again, when the yellow and the dappled bulls were gathered into one herd they stood in such a manner that their number, beginning from one, grew slowly greater till it completed a triangular figure,[30] there being no bulls of other colours in their midst nor none of them lacking.

The description of the groups of bulls – white and black standing firm, equal in depth and breadth (as in a square), contrasted with the yellow and dappled bulls forming a triangle – is visually suggestive both of geometrical shapes and of the sort of figured numbers – here, a rectangular number and a triangular number – that Aristotle had described being created by using pebbles to bring 'numbers into the forms of triangle and square' (*Metaphysics* 1092b12, trans. Ross).[31] In other words, there is the suggestion that finding the square number and the triangular number is an achievable task.

At the end of the statement of the second part of the *Cattle Problem*, the reader is promised:

> If thou art able, O stranger, to find out all these things and gather them together in your mind, giving all the relations, thou shalt depart crowned with glory and knowing that thou hast been adjudged perfect in this species of wisdom.

However, being crowned with glory and judged perfect in this kind of wisdom is no easy feat. The seemingly simple statements of the two parts of the problem belie the surprisingly difficult character of the *Cattle Problem*, the solution of which is a fantastically large number. Scholars specialising in the work of Archimedes, including Reviel Netz (2009a: 34), believe that the problem could not have been solved in antiquity. The modern 'discoverer' of the problem, Gotthold Ephraim Lessing (1729–1781), offered what was an incorrect solution, and a number of other partial solutions were published in the nineteenth century. Modern results depend on computer calculations.[32]

---

[30] Or, a triangular number; see Heath 1912: 319 and 323.

[31] The terms 'figural' or 'figurate' are also used. Imagine three pebbles (or six, or ten, and so on) being laid out to form a triangular plane figure; square numbers can be used to form a square plane figure.

[32] See Lessing 1773: 422–433 for the poem and discussion. Wurm 1830, through an ambiguity in the text, offers a solution to part of the problem. Amthor 1880 discussed the complete problem, and partly solved it. See Heath 1912: 319–326 for discussion of the earlier history of attempts to solve the

For us, there is the question as to why this intriguing and difficult problem was presented as a poem. If the author was Archimedes, we cannot suggest that metre was his primary mode of communication, for his other works are in prose. One motivation for setting the problem in metre might lie in the relationships it allowed Archimedes to highlight between himself and Eratosthenes, and between their interests and those of Homer.[33]

Eratosthenes began his now lost *Geography* (*Geographica*) with a discussion of Homer (fragments survive in, for example, Strabo's *Geography*; see Geus 2002 and Roller 2010); he thought that the Homeric poems could not be used as reliable sources for geography since, in his view, the poet had intended to entertain rather than instruct (Strabo *Geography* 1.2.3; in Roller 2010: 41–42; see also Pfeiffer 1968: 166; Fraser 1972: vol. 1, 526–527). And, in the *Cattle Problem*, there is an important allusion to Homer's *Odyssey* and the cattle of the sun. In the very opening lines of the *Odyssey* (1.6–10), the cattle of Helios (the sun) are mentioned, foreshadowing the forbidden slaughter of these livestock by Odysseus' companions in Book 12. Indeed, the precise number of animals (seven herds of cattle, and of sheep, with fifty in each), and the place where they pasture, the island of Thrinacia, are specified in Book 12 (lines 127–130).[34] There, the goddess Circe speaks to Odysseus, stating: '... you will reach the island Thrinakia, where are pastured the cattle and the fat sheep of the sun god, Helios, seven herds of oxen, and as many beautiful sheepflocks, and fifty to each herd' (trans. Lattimore 1975: 188); Odysseus' landing at Thrinakia and encounter with the pasturing cattle and sheep of Helios had previously been prophesied as part of the testing challenges in his future, in Book 11 (106–109). These references to Homer, Helios, to numbers of herds of cattle and other livestock (including sheep) and to Thrinacia call to mind

---

problem, in which he estimates the space required just to write out the large numbers required for the solution; he suggests it would take 660 pages. Williams, German and Zarnke 1965 published a report of their solution, achieved using two different IBM computers, and printed on forty-two computer sheets. See also Fowler 1980 and Vardi 1998.

[33] Indeed, Knorr (1986: 295) suggested that Eratosthenes composed the first part of the problem, and that the second part is Archimedes' response. See now Netz 2009a, Leventhal 2013 and 2015, and Benson 2014 (who argues that Archimedes was the author of the entire poem) for reappraisals of the poem from a literary standpoint, considering also Archimedes' interest in the geography of his native Sicily.

[34] Thucydides 6.2.2 tells us that Sicily was originally known as *Trinakria*. (Right before this, in relating the early history of Sicily, he opines that '[we] must be content with the legends of the poets, and every one must be left to form his own opinion' [trans. B. Jowett], suggestively linking this name to ancient poetry.) Strabo *Geography* 6.2.1 explains that 'Sicily is triangular in shape; and for this reason it was at first called "Trinacria", though later the name was changed to the more euphonious "Thrinacis." Its shape is defined by three capes', namely Pelorias, Pachynus and Lilybaeum (trans. Jones, 1924). See also Netz 2009a: 167–168, Benson 2014: 189–192.

various associations, including erudite scholarship on epic poetry as well as sophisticated calculation: in addition to his work on geography (which engaged with Homeric poetry), Eratosthenes' interest in number theory is also well attested.[35] By triangulating himself between Eratosthenes (arguably one of the greatest intellectuals of his age) and Homer (revered as one of the greatest Greek poets), Archimedes (if he was the author of the *Cattle Problem*) highlights intellectual bonds amongst the three, via numbers and poetry. By addressing the *Cattle Problem* to Eratosthenes, the author insinuates a close relationship with these seemingly diverse intellectual interests, forging a close link between mathematics and poetry (including metre and myth). In locating the *Cattle Problem* in Thrinacia (Sicily), the home of Archimedes as well as Helios' herds, Homer's detailed attention to the numbers of cattle is recalled and transformed into a challenging mathematical puzzle, worthy of the great mathematician. And there are many features of the *Cattle Problem* that suggest it was intended for a wider audience than simply Eratosthenes and his students. Archimedes and Eratosthenes, or whoever the presenter and intended recipient of this mathematical poem might have been, were not alone in their interest in mathematical problem-poems, as we see from the examples collected in the *Greek Anthology*.

Mathematical Epigrams in the *Greek Anthology*

The *Greek Anthology* (*Anthologia Graeca*) contains poems, mainly epigrams (Cameron 1993: 13), ranging from the Classical to the Byzantine periods. Two manuscripts – the Palatine Anthology (*Anthologia Palatina*) of the tenth century, credited to Constantine Cephalas; and the fourteenth-century Anthology of (Maximus) Planudes, also known as the Planudean Anthology – are the principal sources of the *Greek Anthology*. The *Greek Anthology* contains, among others, more than forty poems which are mathematical problems presented as epigrams; many of these are associated with Metrodorus. While some of the problems may have been his own, he is generally credited not with having devised the problems, but with having collected them (Midonick 1965: 496); it is very likely that the problems would have been posed – and possibly even the epigrams composed – much earlier.[36]

---

[35] See Nicomachus *Introduction to Arithmetic* 1.13 for a description of Eratosthenes' 'sieve', a method for finding prime numbers.
[36] The modern version of the *Greek Anthology* is based on the tenth-century manuscript found in the Palatine library at Heidelberg in 1606 by Claudius Salmasius. Known as the Palatine manuscript, it consisted of fifteen books of short poems or epigrams, organized according to subject matter. For

Book 14 of the *Greek Anthology* is devoted to riddles, oracles and mathematical problems. The mathematical poems are a form of 'story-problem' requiring some level of comfort with mathematics and also expecting a certain degree of familiarity with poetic traditions. Like the *Cattle Problem*, those in the *Greek Anthology* were presented without solutions; the epigrammatic form seems to have been particularly suitable for posing problems. But unlike the *Cattle Problem*, those in the *Greek Anthology* were apparently thought to be soluble, as the evidence of the solutions provided in the *scholia* attests (as edited by Tannery 1895).

Judging from their content, many of the mathematical problems may have originated as early as the fifth century BCE or even earlier. Plato, in the *Laws* (Book 7), presented the Athenian stranger, the Cretan Clinias and the Spartan Megillus discussing, amongst other topics, questions having to do with the proper education of children. The importance of learning arithmetic, counting and calculation – and of having fun while doing so – is stressed, when the Athenian states (at 819a-c) that:

> One ought to declare, then, that the freeborn children should learn as much of these [mathematical] subjects as the innumerable crowd of children in Egypt learn along with their letters.[37] First, as regards *counting*, lessons have been invented for the merest infants to learn, *by way of play and fun*, – modes of dividing up apples and headpieces [hairbands, garlands or crowns], so that the same totals are adjusted to larger and smaller groups, and modes of sorting out boxers and wrestlers, in 'byes' and 'pairs', taking them alternately or consecutively, in their natural order. Moreover, by way of play, the teachers mix together bowls made of gold, bronze, silver and the like, and others distribute them, as I said, by groups of a single kind, adapting the rules of elementary arithmetic to play; and thus they are of service to the pupils for their future tasks of drilling, leading and marching armies, or of household management....[38]

This suggestion of teaching children through play accords with the recommendation in one of Plato's earlier works, the *Republic*, that children's lessons should take the form of play; it is claimed that, while forced exercise

---

the histories of specific anthologies, and the archaeology of their construction, see Cameron 1993; Gow and Page 1965: vol. 1, Introduction and Gow and Page 1968: vol. 1, Introduction.

[37] The Rhind and Moscow papyri give an idea of the state of mathematics in Egypt prior to Plato's time; see Clagett 1999: 113–237.

[38] Trans. Bury 1926: vol. 2, 105, slightly altered and emphasis mine; see note 55 below regarding the translation as 'apples'. The term 'bye' is used in sports (including boxing) to describe 'the position of an individual, who, in consequence of the numbers being odd, is left without a competitor after the rest have been drawn in pairs' (*OED* online: www.oed.com/view/Entry/25545?rskey=UIhP48&result=2#eid; accessed 31 August 2015). The passage in the *Laws* goes on to emphasise the importance of metrology and of geometry.

does the body no harm, nothing learnt by force stays in the mind (Book 7, 536e). Furthermore, educating children through play will also highlight the types of work to which each child is naturally suited.

A number of the mathematical poems collected in the *Greek Anthology* appear to answer the call of the Athenian, dealing with problems related to apples, ornamental headgear and bowls, as well as other objects made of different metals.[39] Some of the poems which seemingly hark back to Plato's suggestions for the education of children are not attributed to Metrodorus;[40] in these we may see deliberate allusions to Plato's writings. For example, in the *Republic* Book 7 (522b-534e), there is extensive discussion of different types of practical mathematics, such as the counting, calculation and geometrical thinking employed by military generals, and theoretical mathematics (including number theory, geometry, stereometry, astronomy and harmonics); mathematics was also a required part of the education of ideal future philosopher-kings. The sort of problems concerned with apples and metal bowls presented in the *Greek Anthology* were understood by some ancient thinkers as belonging to a particular category of mathematics; according to Proclus, such problems were considered to be part of calculation (*logistikē*, distinguished from arithmetic, concerned with the theory of numbers; Proclus *Commentary on the First Book of Euclid's* Elements, ed. Friedlein 1873: 39.7–40.9; trans. Morrow 1970: 32–33). Carl B. Boyer suggested that it was Plato who was responsible for the distinction between arithmetic (in the sense of number theory) and logistic (as computational techniques), for he saw 'logistic as appropriate for the businessman and for the man of war, who "must learn the art of numbers or he will not know how to array his troops". The philosopher, on the other hand, must be an arithmetician "because he has to arise out of the sea of change and lay hold of true being"' (Boyer 1968: 95).[41] The emphasis on calculation stresses the sort of practical problem which might be encountered in everyday life, even when it is couched in the framework of familiar and traditional myth.

---

[39] There are four problems dealing with counting apples; that three are grouped together in the *Anthology* may be significant.
[40] Not all of the mathematical problems found in Book 14 are attributed to Metrodorus. On the education of children, see Scolnicov 2004; Marrou 1956: 69–75 (Plato's recommendations), 142–185 (in Greece), 265–283 (in Rome).
[41] The terminology (for example, of 'logistic') can be confusing, particularly with regard to modern English usage. According to the Scholium to Plato's *Charmides* 165E (Hermann 1878: vol. 6, 290 and translated by Heath 1921/1981: vol. 1, 14–15), problems dealing with the so-called *melite* ('sheep') and *phialite* ('cup') numbers are part of logistic (discussed below), as is the *Cattle Problem*; see also note 54 to this chapter.

Typically, epigrams may make allusions to myth and epic, as ways of aligning with and reshaping tradition. Indeed, many of the mathematical epigrams present story problems which resonate with traditional myth and epic poetry. Take for example the third problem (Epigram 3) in Book 14, dealing with apples.

> Cypris[42] thus addressed Love [Eros], who was looking downcast: 'How my child, hath sorrow fallen on thee?' And he answered: 'The Muses stole and divided among themselves, in different proportions, the apples I was bringing from Helicon, snatching them from my bosom. Clio got the fifth part, and Euterpe the twelfth, but divine Thalia the eighth. Melpomene carried off the twentieth part, and Terpsichore the fourth, and Erato the seventh; Polyhymnia robbed me of thirty apples, and Urania of a hundred and twenty, and Calliope went off with a load of three hundred apples. So I come to thee with lighter hands, bringing these fifty apples that the goddesses left me.'[43]

Here, in a poem about calculating a number – admittedly not as challenging as Archimedes' about the cattle – we find a detailed reference to the Muses, reminding us – again – of the more ancient poets, in this case Hesiod.[44] For, the *Theogony* (76–85) tells us that

> Thus sing the Muses who have their homes on Olympos,
> The nine daughters born of great Zeus,
> Klio, Euterpe, Thalia, Melpomene,
> Terpsichore, Erato, Polyhymnia, Ourania,
> And Kalliope, the most important of all.
> For she keeps the company of reverend kings.
> When the daughters of great Zeus will honor a lord
> Whose lineage is divine, and look upon his birth,
> They distill a sweet dew upon his tongue,
> And from his mouth words flow like honey. (Trans. Lombardo 1993: 63.)

The importance of the Muses, and of poetry, is emphasised. The poetic allusions here engage with texts as central as the Homeric and Hesiodic poems, being linked to other mathematical poems (including another epigram [number 48] dealing with apples and Muses) while emphasising

---

[42] Another name for Aphrodite, born on the island of Cyprus.
[43] 672+280+420+168+840+480+30+120+300+50 = 3360; trans. Paton 1918: vol. 5, 29. This problem is not attributed to Metrodorus. The translations given here from Book 14 of the *Greek Anthology* are from Paton 1918, unless otherwise noted.
[44] There are also resonances with Apollonius Rhodius *Argonautica* 3, 90–160, where Aphrodite admits to Hera and Athene that she has severe difficulties educating her son Eros, and we learn about his games and golden toys, and his travel through the orchard of Zeus. (I thank Jochen Althoff for his insights; personal communication.)

relationships between different types of learning (represented by the Muses themselves). We see here the importance of the Muses, even for doing arithmetic.

Apples are at the core of several of the mathematical epigrams, and there are numerous Greek myths which importantly feature them. For example, Heracles, as the eleventh of his Twelve Labours, was ordered to travel to the Hesperides to fetch the golden apples that had been presented to Zeus following his marriage to Hera; they were guarded by a hundred-headed dragon.[45] Another important story about apples recounts that Eris, the Greek goddess of discord, had not been invited to the wedding of Peleus and Thetis. In a fit of pique, she threw a golden apple inscribed 'for the most beautiful' towards the wedding party; three goddesses each claimed it: Hera, Athena and Aphrodite. Paris of Troy was tasked, by Zeus, with choosing the recipient. While all three goddesses tried to tempt him, Aphrodite promised him with the most beautiful woman in the world – Helen, wife of the King of Sparta. Paris awarded the golden apple to Aphrodite, and got Helen, triggering the Trojan War.[46] And in yet another tale of wooing, Aphrodite played a key role in giving apples to Melanion to throw on the racecourse as he ran against Atalanta; she slowed down to retrieve the apples, lost the race and married him.[47] The reference to Aphrodite as Cypris in the third epigram in Book 14 carries powerful allusions to the significance of apples in traditional stories and myths, particularly those involving love and marriage.[48]

In addition to apples, there are many other elements suggestive of myth, and history (including the account of Herodotus), in these mathematical problems, for example in Epigram 12, with the reference to Croesus (Kroisos):

> Croesus the king dedicated six bowls weighing six minae, each one drachm heavier than the other.[49]

Herodotus (*Histories* 1.50–52) offers a long list of all the items that Croesus sent to the oracle in Delphi; many have the exact weight given.[50] Croesus

---

[45] Apollodorus *Library* 2.5.11 recounts Heracles' labour of fetching the apples.
[46] Hyginus *Fabulae* 92 recounts the story.
[47] Recounted in Hyginus *Fabulae* 185 and Apollodorus *Library* 3.9.2.
[48] Even Plato was credited with having composed epigrammatic love poems referring to meaning-laden fruit; see numbers 7 and 8 in the edition by Cooper and Hutchinson 1997: 1742–1745, on p. 1744. Although these poems are concerned with mathematics, the listener or reader is meant to learn another important lesson.
[49] According to Paton 1918: vol. 5, 33 the weight of the first bowl is 97.5 drachmas.
[50] Jochen Althoff (personal communication) reminded me of this list, behind which lies the story (Herodotus 1. 46–48) of how Croesus tested all the major Greek oracles, with the result that the Delphic was found to be the best.

was the King of Lydia from 560 to 546 BCE until his defeat by the Persians. But by the fifth century at least, J.A.S. Evans has argued, 'Croesus had become a figure of myth, who stood outside the conventional restraints of chronology' (Evans 1978: 36). In the epigrammatic problem here, a mythical setting is given specific mathematical meaning through reference to the practical problems of weighing different bowls, a problem very similar to those mentioned by Plato in the *Laws*. The problems numbered 12 and 50 are reminiscent of references by Plato to the use of bowls made of the same or of different metals as a tool in the education of children (*Laws* 7, 819b-c). For example, Problem 50:

> Throw me in, silversmith, besides the bowl itself, the third of its weight, and the fourth, and the twelfth; and casting them into the furnace stir them, and mixing them all up take out, please, the mass and let it weigh one mina.[51]

While these mathematical problems in the *Greek Anthology* dealing with apples and wealth (be it bowls formed of precious metals, or crowns) resonate with traditional myths, they also refer to potentially real objects and seemingly practical problems. While it is difficult to imagine real-life applications of the complicated problems presented in the poems, the counting of objects like bowls and apples was seen – even by Plato – as something necessary to learn for everyday life. Determining the weight of precious metals was necessary to ascertain their value.

Indeed, it is the character of dealing with things that are sensible and material (in contrast to abstract ideas, such as number) that characterise these problems of calculation, according to Proclus, and the scholiast to Plato's *Charmides* 165 (see Cohen and Drabkin 1948: 4; Klein 1968: 12–13; Hermann 1878: vol. 6, 290 for the scholium).[52] Helpfully, the scholiast notes that

> Logistic is a science, which concerns itself with *counted things*, but not with *numbers*, not handling that number which is truly a number, but positing that which is one [*namely one definite thing*] as the unit itself, and that which is being counted [namely the particular assemblage] as the number itself, so that this science takes, for example, three things as 'three' and ten

---

[51] Solution given by Paton 1918, vol. 5, 51: the bowl weighs $3/5$ of a mina, or 60 drachmae (one mina=100 drachm).

[52] There are other 'mathematical' problems that are eminently practical, referring to instruments which required a certain degree of mathematical understanding to be produced. For example, there are epigrams about sundials; see Evans 2005: 287–288.

things as 'ten'; to these things it applies the theorems of [pure] arithmetic (trans. Klein 1968: 12; his bracketed comments and italics).

The scholiast goes on to explain that logistic 'investigates both that problem which is called the "cattle problem" by Archimedes and also the "sheep-numbers" and "bowl numbers", the latter with reference to bowls, the former with reference to a flock of sheep; and also with respect to other classes of bodies perceptible by the senses, it investigates their multitudes and comments on these as if they were perfect. All countable things are its material' (trans. Klein 1968: 12). Further, the 'parts of *logistikē* (the science of calculation) are the so-called Greek and Egyptian methods in multiplications and divisions, and the additions and subtractions of fractions … The aim of it all is the service of common life and utility for contracts, though it seems to deal with things of sense as if they were perfect or abstract' (trans. Heath 1964: 111). The concern of calculation with perceptible objects that can be counted is also emphasised by Proclus, who provides further explanation: 'nor does the student of calculation consider the properties of number as such, but of numbers as present in sensible objects; and hence he gives them names from the things being numbered, calling them sheep numbers [*mēlitēs*] or cup numbers [*phialitēs*]' (trans. Morrow 1970/1992: 33).[53] The importance of integers is stressed: 'he does not assert, as does the arithmetician, that something is least; nevertheless with respect to any given class he assumes a least [or smallest number], for when he is counting a group of men, one man is his unit' (Proclus, 40.5, trans. Morrow 1970/1992: 33; the parenthetical gloss, 'smallest', is mine). The emphasis on counting and on 'units' is characteristic of logistic: counting is done with reference to units of material and perceptible objects.[54]

The words for 'apple' and 'sheep' are the same (*mēlon*), and since both are used with reference to calculational problems, there is a certain degree of built-in ambiguity as well as some intertextuality (particularly with regard to the meaning 'sheep'), between the epic poets, Plato (particularly in the *Laws*), the references to calculational (logistical) problems, and the

---

[53] See also Morrow 1960: 345–346 on sheep numbers; the sheep being counted may have been toys.
[54] For more on logistic, see Klein 1968: 6–8, and his Part I. Historians of mathematics have considered whether some of the mathematical epigrams show signs of 'algebraic' thinking. Heath 1964: 113 and others – e.g., Cohen and Drabkin 1948: 25 – have described these problems in modern terms, referring, for example, to 'algebra'. Christianidis 1994: 238–240 and 245 regards them as part of 'logistic'. On the relationship between arithmetic and logistic in Plato, see Klein 1968: 17–25 and Hopkins 2011: 154–162. Sesiano 2004: 258 suggests that there likely were other examples of 'recreational' problems.

problems in the *Greek Anthology*, as well as the *Cattle Problem* itself. One can almost imagine that part of the fun was the possible punning involving livestock and fruit.[55]

Problem 49 in Book 14 of the *Greek Anthology* is an example of the application of the so-called 'rule' given by Thymaridas, who may have been an early Pythagorean. Iamblichus, in his third-century CE commentary on Nicomachus' *Introduction to Arithmetic* (first century CE), describes the rule of Thymaridas as a 'flower' or 'bloom';[56] this name seems particularly appropriate in the 'gathering of flowers' (*anthologia*) of the *Greek Anthology* itself.

The problem recalls some of the others we have noticed, with its references to valuable materials:

> Make me a crown weighing sixty minae, mixing gold and brass, and with them tin and much-wrought iron. Let the gold and bronze together form two-thirds, the gold and tin together three-fourths, and the gold and iron three-fifths. Tell me how much gold you must put in, how much brass, how much tin, and how much iron, so as to make the whole crown weigh sixty minae.[57]

This problem, involving the solution of several unknowns, also has mathematical resonances with Archimedes' *Cattle Problem*, but on a different scale.[58] And, as in the *Cattle Problem*, there is also an allusion to myth: the crown may well have belonged to Croesus (who features in Problem 12 of Book 14), or another king. Do we have any sense of who the author of

---

[55] Forms of the word *mēlon* are used in the Homeric and Hesiodic poems to refer to both sheep and apples, e.g., to sheep at Homer *Iliad* 10.485, *Odyssey* 14.105, Hesiod *Works and Days* 786 and to apples at *Odyssey* 7.120 and Hesiod *Theogony* 215 and 335. Morrow 1960: 345, n. 170 argues for the use of the term 'sheep number', while acknowledging that there are numerous readings of 'apples' in Plato's *Laws* 819b-c; cf. Heath 1964: 113, preferring apples, and also the translation by Bury of the passage from the *Laws*, quoted above (at note 38). Heath 1921/1981: vol. 1, 14 while stating a preference for apples, notes that the Greek word *mēlon* means both 'sheep' and 'apple'; he, as does Morrow, acknowledges that the scholiast to Plato's *Charmides* (165E) points out that the work could refer to either (see also note 41 here).

[56] See Thomas 1939–1941: vol. 1, 138–41, for Iamblichus' text. The 'rule' has been understood by Heath 1921/1981: vol. 1, 94, and some other historians of mathematics, to be 'algebraic' in conception, but this interpretation is open to debate. To judge from Iamblichus' account, it does not appear that Thymaridas used symbols. On historiographical issues relevant for the study of Greek mathematics and questions relating to algebra, see, for example, Unguru 1975; Rowe and Unguru 1981.

[57] Trans. Drabkin, in Cohen and Drabkin 1948: 26. The solution given by Paton 1918: vol. 5, 51 is: gold 30½, brass 9½, tin 14½, iron 5½.

[58] As Heath 1921: 94 (=1964: 115) notes, the 'flower' or 'bloom' is obscurely worded; the rule was used to solve 'a certain set of *n* simultaneous simple equations connecting *n* unknown quantities'. And, it 'was evidently well known' (as indicated by its special name). See also Christianidis 1994.

such a mathematical problem-poem may have been? And who would have likely been the audience?

Jens Høyrup has argued that mathematics (at least the mathematics of literate cultures) was not always treated and taught as a formal science by specialists alone; he points to the mathematical riddle collection of Book 14 of the *Greek Anthology* as evidence that mathematics was not 'always or inherently treated as a formal science'. As he notes, the collection in the *Greek Anthology* combines properly arithmetical problems with other matters, including oracles, and he argues that 'such collections ... were not put together by ... professional mathematicians, nor as the basis for systematic mathematics teaching; they represent the view of the more or less informed *outsider*' (Høyrup 2006: 96–97, emphasis in the original). In other words, the mathematical poems in the *Greek Anthology* may not have been devised and written by specialist mathematicians, as the *Cattle Problem* is presumed to have been, but rather by interested non-specialists, knowledgeable about both mathematics and poetry. And it is possible to imagine learned enthusiasts – also interested both in mathematics and in poetry – enjoying these epigrams, engaging in a sort of mathematical recreational activity.[59] Indeed, the composition, hearing or reading and solving of these texts required a number of skills: the solving of the problem would have been one (mathematical) challenge; the setting out of the problem in a poetic epigram (with allusions to other texts and cultural expressions, including poetry, myth and mathematics), and the informed multilayered 'reading' of these epigrams would have presented others. One can well imagine that this is just the sort of 'party-piece' that would have won favour amongst some symposiasts as an intellectually stimulating and pleasurable game; riddling poems of various kinds were a standard form of sympotic literature, whether actually delivered at symposia or composed and collected as the kind of thing which might be offered in that setting.[60] These mathematical epigrams were not just for 'specialists', even though some of the problems are sophisticated and challenging; their survival in the *Greek Anthology* implies that many readers with a range of interests would have been intrigued by them. Their reliance on many commonplaces of ancient life reinforces their apparent accessibility and appeal.

---

[59] On mathematical 'recreational' problems and riddles, see Høyrup 2001: 87–88.
[60] See West 2012 on riddling as a standard form. Bing 1998: 32–33, n. 37 suggests that it is unlikely that epigrams were composed during the symposium itself.

Practical Problems and Riddles about Life

In the next chapter, we will look closely at a letter attributed to Eratosthenes, familiar to us as the named addressee of the *Cattle Problem*. Early in the letter, the author offers a problem he attributes to an unnamed tragic poet, one of the most famous mathematical problems of Greek antiquity: how to double a cube. In the account of the unnamed poet, the problem is posed by the necessity to increase the size of a mausoleum constructed for the dead son of King Minos.[61] As we have seen, the mathematical problems presented as epigrammatic poems similarly address practical problems, on first reading: counting cattle, apples and bowls. These epigrams also reflect long poetic traditions, including those of epic. Furthermore, the mathematical epigrams set out the problems as riddles; this is key to their appeal and function, which relate to other aspects of Greek culture in which riddling featured.

So, for example, as mentioned earlier, a version of the Sphinx's Riddle, about the stages of life and which involves numbers, also appears in the *Greek Anthology*, in Book 14 (number 64), alongside other riddles and questions about men's lives; questions about the length of individuals' lives are raised in epigrams 126 (on the length of Diophantus' life) and 127 (on Demochares), both calling to mind Herodotus' report of Solon's numerical description to King Croesus of the length of a man's life (*Histories* 1.29–32, esp. 32; a portion of a poem by Solon in West 1993: 80, fragment 27, mentions the ten seven-year stages of life). The slightly different versions of the Riddle of the Sphinx point to a degree of fluidity and instability in the riddle itself, before it was preserved as a written text. This fluidity and instability points to another possible motivation for committing the mathematical puzzles to writing, even if they were first designed as entertainment for a symposium: to preserve the numbers, which are crucial to the riddle itself (and might easily be misremembered).

Ewen Bowie has pointed to a number of assumptions sometimes made by scholars working on epigrams: that some 'first saw the light of day in a recitation at a symposium and others in a written copy sent to a friend (addressed in the vocative as if present); that poems which had begun life in either medium might be resurrected, possibly many times, in the other; and that all or most poems which had initially been 'published' in these ways were gathered by their poets into collections'. These assumptions are not universally accepted, but they are, nevertheless, helpful to us as we

---

[61] Perhaps unsurprisingly, the *Letter* also figures Plato as a source of mathematical learning.

consider the various ways in which mathematical epigrams may have been encountered by ancient audiences (Bowie 2007: 109).

The number of mathematical poems that survive suggests that such poetry was not simply the preserve of the most accomplished intellectuals, such as Eratosthenes and Archimedes, known correspondents, sending each other challenging problems. Markus Asper, building on the work of Jens Høyrup, has suggested that the riddle was a genre typically used by groups of mathematical practitioners in competitive struggles, in which experts challenged one another. For example, it is not clear how or when one might be confronted with the practical task of doubling a cube, except in a legend. He suggests that 'such riddles are characteristically compound problems, and apply practical methods to improbable problems that already touch upon a theoretical realm' (Asper 2009: 122, citing Høyrup 1997: 71–72). On Asper's reading, the riddle may offer a model for the transition from mathematical practice to theory, as well as something for historians of mathematics to ponder in their accounts of ancient Greek mathematics (Asper 2009: 122–123). But the problems in the *Greek Anthology* seem to be of a different sort, more likely to have been designed to show various types of knowledge, cultural as well as numerical, to have displayed wit, and a wish to entertain and amuse.

Archimedes' *Cattle Problem* and the mathematical problem-poems from the *Greek Anthology* all presume delight in the presentation of seemingly *practical*, possibly 'everyday' mathematical problems proffered in the form of epigrammatic riddles. A range of literary interests and cultural allusions is reflected even in these texts which are sometimes classified rather narrowly as 'mathematical'. These mathematical epigrams point to the pleasures of using poetry for a variety of purposes – being amusing while showing breadth of education and culture. The evidence of these mathematical problem-poems presented in epigrams suggests that mathematics and poetry were – for some authors and readers – not as separate as we may sometimes believe. In fact, the genre of epigrammatic poetry provided an especially apt format for the function of presenting puzzling riddles. If the mathematical epigrams were composed and collected as part of the literature associated with the symposium, they should be understood not simply as intriguing curiosities, but as being deeply rooted within characteristically Greek cultural institutions and practice, which included philosophy, mathematics, mythology, poetry and other forms of art and expression. That mathematics and other technical and scientific topics were the subject-matter of poetry underscores the primary cultural import of poetic forms for ancient Greek, and Roman, culture.

2

Letter

Artemon, the editor of Aristotle's *Letters*, says that a letter ought to be written in the same manner as a dialogue, a letter being regarded by him as one of the two sides of a dialogue. There is perhaps some truth in what he says, but not the whole truth. The letter should be a little more studied than the dialogue, since the latter reproduces an extemporary utterance, while the former is committed to writing and is (in a way) sent as a gift.
[Demetrius] *On Style* 4.223–224, trans. W. Rhys Roberts

Dionysodorus (for I will not withhold this outstanding instance of Greek folly) has a different creed. He belonged to Melos, and was a celebrated geometrician; his old age came to its term in his native place; his female relations who were his heirs escorted his obsequies. It is said that while these women on the following days were carrying out the due rites they found in the tomb a letter signed with his name and addressed to those on earth, which stated that he had passed from his tomb to the bottom of the earth and that it was a distance of 42,000 stades. Geometricians were forthcoming who construed this to mean that the letter had been sent from the centre of the earth's globe, which was the longest space downward from the surface and was also the centre of the sphere. From this the calculation followed that led them to pronounce the circumference of the globe to be 252,000 stades.
Pliny the Elder *Natural History* 2.112, trans. H. Rackham (1: 373)

In antiquity – as today – various sorts of letters were written and circulated; not all letters were intended for private communication. Some, like the letters published in modern newspapers, were clearly meant for a wider readership. Letters were used for a variety of purposes in the Greco-Roman world, including sharing scientific and mathematical ideas and methods. For example, the three letters of Epicurus (born Samos, 341 BCE; died Athens, 270 BCE) which are preserved in the 'Life of Epicurus' by Diogenes Laertius were intended to give philosophical advice and instruction, and are crucial for understanding ancient Epicurean ideas about nature. Other letters presented technical or scholarly work on mathematical, mechanical and medical topics. A number of letters written by ancient Greek mathematicians survives, indicating that letter-writing was a useful mode of communication for them; for example, Eratosthenes, who lived in Alexandria, was the recipient of letters from Archimedes, living in Syracuse.[1]

While other materials were also employed, letters were usually written on papyrus, which was then rolled up and tied with a thread; delivery could be a complicated process. Most private letters relied on merchants and travellers for dispatch, but those with adequate wealth could afford to send a slave. As Patricia Rosenmeyer has explained, in the Classical period, letter-writing was a relatively unusual activity, serving as a symbol of power and authority reserved for the few. As literacy developed, particularly at the more educated levels of society, 'letter writing became part of everyday life in Hellenistic Egypt, both in administrative affairs and in private households' (Rosenmeyer 2001: 21; cf. Harris 1989: 127–128. See also Trapp 2003: 1–42). (It is worth noting that papyrus survived well in Upper Egyptian conditions.)

Letters are texts which are intended to be shared between individuals or members of a community; their format and paratextual markers (naming the addressee as well as the author) normally explicitly state the intended audience or recipient. In this way, relationships between authors and addressees are highlighted; membership in a specific community (religious, intellectual and other types) is reinforced. In some cases the community is a correspondence network in which letters are exchanged rather than simply conveyed from one individual to others. Letters were an especially important form of communication for early Christians; some New Testament texts are presented in the guise of letters.

And, even if their authorship cannot always be guaranteed, there are a number of letters attributed to famous individuals, including philosophers and mathematicians, surviving from antiquity. Artemon of Cassandreia (perhaps second or first century BCE) is credited with publishing Aristotle's correspondence (Trapp 2012), but most of the letters attributed to Plato are thought to be spurious (see Burnyeat and Frede 2015; see also Trapp 2003: 27–31 on pseudepigrapha). In fact, as a genre, letters are

---

[1] The *Cattle Problem* attributed to Archimedes was addressed to Eratosthenes; see Chapter 1.

particularly liable to suspicion regarding authenticity. Some scholars have argued that letters are particularly easy to forge; furthermore, the rhetorical schools used the composition of letters by famous people as a standard exercise (Morrow 1962: 3–5; see also Huffman 2005: 42–43). During the Hellenistic period, libraries were actively collecting Greek material, and there was a market for forgeries.

The letter does not appear to modern readers to have been a particularly characteristic genre of communication for natural philosophers and mathematicians in antiquity, as it was for the early Christian community, even though it has been of great service in facilitating the exchange and transmission of ideas and information on scientific and mathematical topics in more modern periods.[2] Nevertheless, a number of significant letters on scientific and mathematical topics survive, attributed to famous authors, including Epicurus and Archimedes.

Scholars from a range of fields – including anthropology and sociology, as well as literary studies – have pointed to the role of the letter functioning as a gift in various gift exchange societies and situations (Wilcox 2012: 3–22). Ancient Greek and Roman texts presented in the form of the letter had numerous functions, including conducting business, offering advice or consolation, and making public statements. Several of the surviving ancient Greek letters on scientific and mathematical topics seem to have served at least partly as a sort of gift. It is worth looking at those examples more closely, not least because of their importance to history of science. The letters attributed to Epicurus, presenting his natural philosophy, deserve attention especially because they are amongst the only extant texts ascribed to him; whether the epistolary format played a role in their survival is an intriguing question. The *Letter to King Ptolemy* attributed to Eratosthenes of Cyrene provides an especially fine example of the richness of the genre and serves here as a special case study, as an example of an elaborate gift presented in a seemingly simple package – the genre of letter itself.

Epicurus' Letters on Natural Philosophy

Two letters attributed to Epicurus and preserved in their entirety in the biography of Epicurus written by Diogenes Laertius (probably first half

---

[2] Tycho Brahe (1546–1601) and Charles Darwin (1809–1882) are both examples of scientists for whom letter-writing and correspondence networks played crucial roles. See Mosley 2007 and the *Darwin Correspondence Project* (www.darwinproject.ac.uk/).

of the third century CE) in his *Lives of Eminent Philosophers* are particularly important for our knowledge of Epicurean cosmology and meteorology: the *Letter to Herodotus* and the *Letter to Pythocles*.³ The letters of Epicurus are rather brief, intended to serve as *aides-mémoire* for his followers, and can be understood as 'didactic' letters.⁴ Each of these texts is referred to as an *epistolē* by Diogenes Laertius (e.g., 10.28, 10.34, 10.83, 10.121), and *epistolē* is the word used by Epicurus in his reply to Pythocles. The letters structurally have the paratextual features, particularly the opening salutation ('Epicurus to Herodotus, greetings' (*chairein*); 'Epicurus to Pythocles, greetings') that we would expect. At the beginning of the *Letter to Pythocles*, Epicurus makes clear that the two have been corresponding, and indicates how letters have been passed between them:

> Epicurus to Pythocles, greetings.
> In your letter to me, of which Cleon was the bearer, you continue to show me affection which I have merited by my devotion to you, and you try, not without success, to recall the considerations which make for a happy life. (Diogenes Laertius, 10.84, trans. Hicks 1925: vol. 2, 613.)

The letter to Pythocles is personalised and reflects their relationship, while the letter to Herodotus indicates that Epicurus wrote this epitome (as he refers to it: *epitomē*) for 'those who are unable to study carefully all my physical writings' (10.35, trans. Hicks 1925: vol. 2, 565).

The letter format employed by Epicurus was intended to convey his ideas to a broad and literate audience; this audience included those who did not have specialised philosophical interests. Epicurus' school in Athens, the Garden, included women and slaves as members, indicating his aim to direct his philosophy towards a wide range of people.⁵ And, in antiquity, Latin texts as well as Greek were used to communicate Epicurean ideas in a range of different cultural contexts.

The different specific formats of surviving texts associated with Epicureanism indicate the intention to communicate widely, as do explicit statements contained within those texts. For example, in the *Letter to Herodotus* (10.82–83) Epicurus makes clear his aim to present a summary of the chief doctrines of his physics: 'so that, if this statement be accurately

---

³ A third letter is preserved there as well, but is less important for the topic considered here.
⁴ I adopt the term 'didactic' letter from Mansfeld 1999: 5. Translations of Epicurus' letters here are from Hicks 1925: vol. 2.
   In addition to these texts, fragments of a massive Greek inscription (about 80 meters long) erected by Diogenes of Oinoanda (probably second century CE) also provide information on Epicurean teachings; portions of the inscription dealt with Epicurean ideas about physics and astronomy. See M.F. Smith 1993, 1996 and 2003.
⁵ See Clay 2009: 18–22 on the Epicurean use of letters more generally.

retained and take effect, a man will, I make no doubt, be incomparably better equipped than his fellows, even if he should never go into all the exact details'. A key goal of Epicurean philosophy is the attainment of *ataraxia* (freedom from anxiety); once we understand that, in principle, potentially frightening phenomena can be explained naturally, without recourse to the gods, worry about those phenomena is then unnecessary. And while Epicurus indicates that he had developed his views in detail, in the *Letter to Herodotus* he intends that 'the summary itself, if borne in mind, will be of constant service'. Likewise, the *Letter to Pythocles* is also proffered as an *aide-mémoire*, in response to a request from the addressee (10.84): 'To aid your memory you ask me for a clear and concise statement respecting meteorological phenomena; for what we have written on this subject elsewhere is, you tell me, hard to remember, although you have my books constantly with you'. At several points in these letters Epicurus suggests that more detailed explanations are not always more helpful and they may even be anxiety-producing. Yet, he does not exclude those who are more learned from also having the benefit of his letters, outlining his ideas.

Two key aims of Epicurean philosophy – to communicate broadly and to alleviate anxiety – are at the forefront of his cosmology and his explanations of meteorological phenomena. Sufficient understanding of cosmology and meteorology is available to ordinary people to alleviate anxiety about the world and potentially frightening phenomena – such as lightning or thunder – simply through employing common everyday techniques, including using clear language, observations, and analogies to what is already familiar. The presentation of an 'in principle' explanation in a brief format is enough to enable his followers to achieve *ataraxia* with regard to a number of possible fears. Indeed, Epicurus cautions against gaining too detailed knowledge about phenomena: such knowledge may lead to further anxiety and fail to contribute to peace of mind. In his view, astronomical knowledge cannot contribute to happiness; for this reason, he does not advocate the detailed mathematical study of the motions of the heavenly bodies. He explains: '[W]hen we come to subjects for special inquiry, there is nothing in the knowledge of risings and settings and solstices and eclipses and all kindred subjects that contributes to our happiness'. He argues that

> those who are well-informed about such matters and yet are ignorant what the heavenly bodies really are, and what are the most important causes of phenomena, feel quite as much fear as those who have no such special information – nay, perhaps even greater fear, when the curiosity excited by

this additional knowledge cannot find a solution or understand the subordination of these phenomena to the highest causes (*Letter to Herodotus* 79; trans. Hicks 1925: vol. 2, 609).

The brief letter format is ideal for communicating the essence of Epicurus' ideas; from his point of view, that essence is sufficient; detailed knowledge is not only not beneficial but may actually be harmful to one's peace of mind. In the case of communicating his philosophy, for Epicurus, less is more; the letter provides an excellent short format for conveying sufficient information to achieve the goal: freedom from worry. In contrast, in Eratosthenes' *Letter to King Ptolemy*, the relatively – even deceptively – simple letter format provides an opportunity to present an extravagantly rich text in which, arguably, more is more, resulting in a suitably impressive gift for a royal patron.

The *Letter to King Ptolemy*

Eratosthenes of Cyrene (*c*. 285–194 BCE) is credited with having devised a mechanical solution to the problem of the duplication of the cube.[6] His achievement is reported in a letter addressed to King Ptolemy (possibly Ptolemy III, 'Euergetes', or the 'Benefactor'), preserved in Eutocius of Ascalon's (480–540 CE) commentary on Archimedes' *Sphere and Cylinder*.[7] The text has been of great interest to historians of mathematics, as it provides a proof for one of the three 'classic' problems of ancient Greek mathematics, the duplication of the cube (also known as the 'Delian' problem).[8] Individual sections of the *Letter*, for example the geometrical proof, are often the focus of study.[9]

Here I consider the text of the *Letter to King Ptolemy* as it is presented in Eutocius as a whole, coherent text: that is, as a letter concerned with a mathematical problem. However, it is crucial to note that the *Letter* incorporates several other genres and styles of communication, and is an unusually rich text. Reviel Netz includes this text as an example of what he

---

[6] See Folkerts 2000 on the 'mesolabion', the instrument devised by Eratosthenes.
[7] Knorr 1989: 77–78 has referred to Eutocius' collection of solutions to the problem of duplicating the cube as an 'anthology'. He has suggested (144–145) that the *Letter* was dedicated to the fourth King Ptolemy (Philopator); see note 16 to this chapter.
[8] The other two 'classic' problems are the squaring of the circle and the trisection of an angle.
[9] For example, Thomas 1939–1941 divided the *Letter* into two portions in his *Greek Mathematical Works*, in the section on the duplication of the cube. The first portion (ed. Heiberg 1915: 88.4–90.13) is under the heading 'General' (1: 256–261), and the second (90.30–96.27) is labelled 'The Solution of Eratosthenes' (1: 290–297). Thomas does not treat the letter as a single, unified text, the way it was presented by Eutocius.

calls a 'ludic mathematical treatise', that is, 'a work based on obtaining results in surprising, intricate ways, where the author brings out his own voice in rich, modulated ways, and where the textual surface is often made deliberately opaque by, say, long passages of calculation'.[10] I agree that the text is full of surprises and intricacies, and I concentrate here on the reasons for choosing the letter as a format for communicating a solution to a mathematical problem. Further, it must be noted that the authenticity of the *Letter* has been discussed at length.[11] However, by treating the text as a whole, and recognising its character as an elaborate letter that incorporates different modes of expression, the richness of the text is accentuated and the question of Eratosthenes' authorship becomes less problematic. This text, which has attracted so much interest from historians of mathematics, is a particularly appropriate case study in our consideration of ancient scientific and mathematical letters, for it demonstrates the potential power of the genre.

The *Letter to King Ptolemy* contains several layers of meaning and communication. Here, the mathematical character of the *Letter* will be discussed only briefly, as will particular literary features.[12] There are two distinct pieces of poetry contained within the letter. The first is, supposedly, a quotation from an ancient tragedian; the second is a dedicatory epigram, presented as having been written by the author of the letter and inscribed on a monument announcing the author's discovery of his solution. We might also ask why the author wrote a letter, describing his discovery, if he had already erected a monument for the same purpose.

Eratosthenes was tutor to Ptolemy III's son Philopator ('Father-loving'), and succeeded Apollonius Rhodius as head of the Alexandrian Library. In his own time Eratosthenes was regarded as an accomplished and versatile scholar, as well as a poet. His calculation of the circumference of the earth (252,000 stades) earned him renown (see Geus 2002: 223–259; Aujac 2001; Fraser 1972: vol. 1, 413–415); even in our time it is viewed as remarkably

---

[10] Netz 2009a: 108 contrasts the ludic style to two others: the survey and the pedagogical.
[11] On stylistic grounds, Wilamowitz-Moellendorff 1894/1971: 48–70 thought that the *Letter to King Ptolemy* was a forgery, and many historians of mathematics have accepted his verdict. See also Knorr 1986: 42, n. 10, commenting on Heiberg 1915: vol. 3, 89 note; Heath 1921/1981: vol. 1, 244–245; Thomas 1939–1941: vol. 1, 256 note a; van der Waerden 1954: 160. Some regard the letter itself to be inauthentic, but have taken portions of the *contents* (namely the epigram and the descriptive text directly preceding it) to be genuine. For the purposes here (attempting to understand the text as an example of a letter used to communicate information of mathematical interest), the issue of authenticity is not crucial; the interest of the letter as a text is not diminished even if it is the work of a literary forger of the period. Nevertheless, the question of authenticity will be returned to later.
[12] See Taub 2008b for a more detailed discussion.

accurate. In addition to his interests in mathematical and descriptive geography, he is known to have worked on a number of mathematical problems.

The letter attributed to Eratosthenes is not addressed to Archimedes, or even another scholar, but to his ruler and royal patron, Ptolemy III. Intellectual patronage was a hallmark of the Hellenistic monarchs, especially the Macedonian kings of Egypt, the Ptolemies. Indeed, it was Ptolemy I or II who established the Museum and Library at Alexandria, where eventually Eratosthenes became the third Librarian (see Fraser 1972: vol. 1, 305–335). The production of creative work, offered as gifts, for their supporting patrons was an important duty of intellectuals within the Greco-Roman world. The *Letter to King Ptolemy* credited to Eratosthenes is a particularly appropriate gift: an elaborately crafted mathematical-*cum*-literary text which demonstrates the author's – that is, the Librarian of Alexandria's – desire and ability to honour and amuse his royal patron.

The *Letter to King Ptolemy* celebrates Eratosthenes' solution to the problem of the duplication of the cube. While the *Letter to King Ptolemy* has both a formal beginning (a salutation) and a formal ending (a *sphragis*, or 'seal', which closes the epigram) which both associate Eratosthenes with the text, very little other material contained there contextualises its composition and communication. That Eratosthenes was interested in the solution to the problem of the doubling of the cube is attested by two other texts, which preserve fragments from his work known as the *Platonicus*, one by Theon of Smyrna (*fl. ca.* 115–40 CE), the other by Plutarch (born before 50 CE, died after 120 CE). Scholars have argued about the significance of the apparent contradictions between the *Platonicus* and the *Letter* (see Huffman 2005: 370–385 for an overview). While a comparison of the two is not possible here, Eratosthenes' interest in the problem is well attested, and supports the view that he was the author of the *Letter*.

Eratosthenes does not simply offer a statement and proof of the problem of doubling the cube. The text transmitted by Eutocius is lengthy (more than five pages in the modern edition by J.L. Heiberg) and, from a formal standpoint, rather complicated, while being bounded by the structures of a letter. The text begins with a salutation: 'To King Ptolemy Eratosthenes sends greetings', a typical opening for a letter. This is followed by an historical introduction, setting out the background to the announcement of Eratosthenes' solution to the problem. The author is at pains to contextualise his achievement, recounting at some length the history of various attempts to solve the problem, which were motivated, he explains, not only by an interest in mathematical puzzles but also by the desire to obey an oracle prescribing the doubling of an altar.[13] The author of the letter recounts a story of an ancient mathematical puzzle, quoting a riddling poem from an unnamed ancient tragedian, in which King Minos was concerned with the preparation of a tomb for his son Glaucus; told that the tomb would be a hundred feet each way, he requested that it be doubled, by doubling each side. The author of the *Letter* – pointing to the mathematical interest of the problem – explains that this doubling of the sides would not double the size of the overall tomb: 'for when the sides are doubled, the surface becomes four times as great and the solid eight times'. He then recounts that 'it became a subject of inquiry among geometers in what manner one might double the given solid, while it remained the same shape, and this problem was called the duplication of the cube; for, given a cube, they sought to double it' (trans. Thomas 1939–1941: vol. 1, 259; Knorr 1986: 23 discusses the problem).

Offering an account of previous attempts to solve the problem, the author of the *Letter* makes specific reference to several people, including Hippocrates of Chios, who is here credited with having been the first to discover 'that if a way can be found to construct two mean proportionals in continued proportion between two given straight lines, the greater of which is double the lesser, the cube will be doubled'. But the author notes that 'his difficulty was resolved into another no less perplexing' (trans. Drabkin 1948: 63). A legend about the Delians is then recounted: endeavouring to obey an oracular order to double an altar, they had difficulties and sent a special team 'to ask the geometers of Plato's school in the Academy to find the solution for them'. The efforts of Archytas of Tarentum, Eudoxus, and Menaechmus – all famous in their own right – are then duly noted.

The author of the *Letter* then announces – in the first person – his discovery of 'an easy method' for finding the mean proportionals between two given lines. He explains that 'with this discovery we shall be able to convert into a cube any given solid whose surfaces are parallelograms, or to change it from one form to another, and, again, to construct a solid of the same form as the given solid but larger, i.e., preserving the similarity' (trans. Drabkin 1948: 64). He proudly points to the value of this method for constructing altars and temples (reminiscent of the original

---

[13] Knorr 1986: 24 and 39–40 discounts the external motivation for the problem. Huffman 2005: 378 thinks it likely that developments external, as well as internal, to mathematics may have played a role.

dilemma cited by the unnamed tragedian, as well as that faced by the Delians), converting liquid and dry measures, and even for military applications, such as increasing the size of catapults.[14] (This latter reference to military uses will recall, to some readers, the achievements of another of Erastosthenes' correspondents, Archimedes, who was renowned in antiquity for his work on war machines, including the catapult. The author of the *Letter to King Ptolemy* thus implies that Eratosthenes is a competitor to Archimedes; the *Cattle Problem*, attributed to Archimedes and addressed to Eratosthenes, reinforces the sense of competition between letter-writing mathematicians.)

Turning then to explain his achievement, the author of the *Letter to King Ptolemy* states that he has set out the demonstration and the construction of his instrument. The demonstration is in the form of a formal proof, which is then followed by a description of physical details pertaining to the actual construction and use of the instrument – referred to by others, but not in the *Letter*, as the 'mesolabe' or 'means-taker', the moving ruler used in the mechanical solution.[15] The *Letter* closes with a detailed description of the monument set up by Eratosthenes to commemorate his solution of the problem, conveyed in an epigram, which also serves as the signature, or seal, to the letter.

Layers of Communication and Meaning

The *Letter to King Ptolemy* contains different sections that relate specific types of information: the greeting (identifying the 'author' and 'recipient'),[16] an historical account, geometrical proof, and signature (in this case, identifying both the recipient and the author). The *Letter*, while adhering to an epistolary structure, also contains (or reflects) several other genres of communication. The precise character of the structures (some of which are paratextual, such as the greeting) and genres is open to interpretation, but elements of the following can be found: (1) salutation; (2) quotation from tragedian; (3) *historia*; (4) *mythos* (the Delian legend); (5) geometrical proof (two versions, one of which is mechanical); (6) instructions (for

---

[14] These practical tasks resonate with some of the uses of mathematical learning mentioned in Plato's *Laws* (Book 7, 819a-c; see Chapter 1).

[15] See, for example, Folkerts 2000.

[16] Knorr 1989: 144–145 has suggested that the *Letter* was dedicated to the fourth King Ptolemy (Philopator), Eratosthenes' tutee, perhaps on the occasion of the endowment of royal honours on the infant heir apparent, the fifth Ptolemy (Epiphanes); on this reading, the *Letter* would have been written late in Eratosthenes' career. See also Wilamowitz-Moellendorff 1894/1971: 65–66 on ambiguities in the epigram.

using the 'mesolabe' instrument); and (7) epigram, which also functions as a 'seal' (*sphragis*) or 'signature' here. Finally, the description of the monument points to another important form of communication, a form that is especially public, costly, and intended to be lasting.

Because the *Letter to King Ptolemy* contains a number of genres and styles of communication, it could be argued that the *Letter* is not simply a 'letter'; rather, the letter format almost provides an 'envelope' for a text which embraces several other genres and styles of writing, including poetry, as well as technical mathematical discourse. The rich character of the *Letter to King Ptolemy* suggests that the text was intended to convey more than just a solution to the problem of the duplication of the cube. The author of this text – concentrating on the duplication of the cube – engaged with several different literary genres, including the letter itself; while the text includes certain features particularly associated with mathematical texts, such as technical language, diagrams, and the use of formal proof or demonstration, there is no indication that the text was intended for a narrow or homogeneous audience.

The *Letter to King Ptolemy* operates on several levels, offering, as it were, something for everyone – those who enjoy mathematics, as well as those who prefer poetry.[17] The variety of styles and literary forms contained in the *Letter* may have been included with the aim of entertaining the royal patron and other readers. The different genres contained in the *Letter* accomplish different communicative tasks, and convey – even indirectly – different messages and information. That the formal proof is a key to the significance of the entire text is emphasised by its placement in the centre; for some readers, the geometrical proof may have been rendered more interesting and understandable by having been set within an historical context by the author of the *Letter*.

Given that the *Letter* is addressed to a royal patron, the range of registers may seem, at first, surprising. But a communication dedicated to royalty can also operate for different audiences: encountering the letter serves to make the reader feel like a 'king for a day', at the same time ensuring that the offering is 'fit for a king'. The royal association of various sorts of texts in antiquity – including letters, pharmaceutical recipes and

---

[17] Aristotle recognised that for some styles of argument and presentation, audiences have clear expectations: 'Some people do not listen to a speaker unless he speaks mathematically, others unless he gives instances, while others expect him to cite a poet as witness' (*Metaphysics* 995a5-8, trans. Ross). Cuomo 2001: 32 has noted that mathematics was associated 'not only with a certain subject-matter (numbers, geometrical figures), but also with a certain style'.

books – seems to have been adopted as a tactic to attract attention, while also enhancing the author's reputation (see Totelin 2004).

Considering the text as a letter, Ulrich von Wilamowitz-Moellendorff was critical of the style of the *Letter to King Ptolemy* on several grounds, and based his argument against its authenticity on these stylistic considerations. In his opinion, the letter lacked a sense of style as well as appropriate recognition of its occasion. Wilamowitz concluded that Eratosthenes, a man renowned for his literary achievement, could not possibly have written such a pedestrian letter to his royal patron. In his view, the letter was an invention of late antiquity, cobbled together by someone who had seen the votive monument described therein, and crafted the *Letter* as an accompanying text. In Wilamowitz' view, the epigram which closes the text is genuine, but the *Letter* itself is not (Wilamowitz-Moellendorff 1894/1971).

Wilamowitz further argued against the authenticity of the *Letter* on the ground that it did not share the cultivated style of the period used for private letters, dedications of books and also scientific texts; he mentions specifically a letter from Apollonius of Perga and Archimedes' *Sand-Reckoner* (Wilamowitz-Moellendorff 1894/1971: 53). But, in fact, the opening of *Letter to King Ptolemy* is very similar to that found in letters and other texts by mathematicians of the period, even those addressed to a royal patron. For example, Archimedes opened his *On the Sphere and the Cylinder* with the following salutation: 'Archimedes to Dositheus, greetings' (*chairein*) (see Netz 2004: 31); Apollonius of Perga similarly began his *Conics* (1.2): 'Apollonius to Eudemus, greetings' (for the opening to the *Conics*, see Heath 1896: lxix). Wilbur Knorr argued strongly against Wilamowitz' appraisal of the letter, basing his criticism of Wilamowitz' conclusion on the conviction that the *genre* of communication is important. The heart of Knorr's criticism was of what he regarded as Wilamowitz' 'failure to reckon with the character of the appropriate literary genre'; he implies that Wilamowitz had not taken the genre of 'letter' seriously enough.[18]

---

[18] Knorr 1986: 17–24, quotation on 20. In 1986 Knorr emphasised that the text should be treated as a letter; in 1989: 144 he argued that Eratosthenes' letter 'falls neatly within the genre of mechanical writing in [his] time'. In Knorr's view, the *Letter to King Ptolemy* should not be regarded as inauthentic.

A number of scholars regard both the letter and the epigram as having been written by Eratosthenes. Knorr 1986: 19–22 argued against Wilamowitz's rejection of the authenticity of the letter on several grounds, not least that Wilamowitz's 'forger' had not limited himself only to what appeared in the epigram, but went beyond it in a manner that indicated real interest in the subject. He argued the letter's 'banality' was 'the strongest indication of the letter's authenticity'; in his

Countering Wilamowitz' criticisms, Knorr argued that Archimedes' *Sand-Reckoner* 'provides a good parallel' to the *Letter to King Ptolemy*; Carl Huffman noted that both 'start with very plain addresses to the king'.[19] Both texts are framed as communications to royal patrons (king Gelon II, in the case of Archimedes'); both begin rather simply, without flourish.[20] Archimedes and Eratosthenes, after their brief acknowledgements of the king, launch into discussions of their mathematical interests. In contrast, Apollonius adopts a more personal note with his correspondent Eudemus, enquiring after his health and stating that he himself is well (*Conics* 1.2; see also Heath ed. 1896: lxxii–lxxv for examples of Apollonius' letters opening further books of the *Conics*). This contrast may indicate a difference between public versus private communications: those addressed to royal rulers may have been intended for publication to a wider audience.[21]

Indeed, one ancient author (probably late Hellenistic or early Roman period), concerned with epistolary style, suggested in *On Style* (223) that the preferred style for a letter is rather plain. The author notes (234) that 'since occasionally we write to States or royal personages, such letters must be composed in a slightly heightened tone. It is right to have regard to the person to whom the letter is addressed'. But, the author warns that 'the heightening should not, however, be carried so far that we have a treatise [*sungramma*] in place of a letter, as is the case with those of Aristotle to Alexander and with that of Plato to Dion's friends'. He concludes (235) by noting that 'from the point of view of expression, the letter should be a compound of these two styles, viz. the graceful and the plain' (trans. Roberts 1902: 173 and 177; on Demetrius, to whom *On Style* is often credited, see Russell 2012).

---

estimation, 'surely a forger would have let his imagination run wider'; see also Knorr 1989, particularly chap. 6, in which he considered how literary aspects, such as terminology and style, illuminate the question of authenticity. Netz 2002: 213 n. 57 follows Knorr 1989 in considering the entire text genuine; Zhmud 1998: 216, Geus 2002: 195 also regard the *Letter* as genuine. Huffman 2005: 376 noted that in the body of the letter, the account of solutions offered by other people is more precise than that provided in the epigram; in his view, this argues against Wilamowitz' suggestion that the epigram was genuine but not the letter, as this greater degree of precision could not have been derived from the epigram alone.

[19] Knorr 1986: 20; Huffman 2005: 374. Frances Willmoth has suggested (personal communcation) that the rather simple salutation may reflect an already established patronage relationship.

[20] The *Sand-Reckoner* does not have a salutation, but is nevertheless addressed to Gelon, beginning: 'There are some, king Gelon, who think that the number of the sand is infinite in multitude…' (trans. Heath 1912: 221). Some mathematical texts begin with a letter-type salutation, even though other epistolary conventions, such as a formal signing-off, may not be present.

[21] The simple salutation offered to Gelon by Archimedes contrasts strongly with the effusive greeting proffered to the Emperor Titus in Pliny the Elder's epistolary preface to his *Natural History*. (The *Natural History* is the subject of Chapter 3.)

The Choice of a Letter

Why was this particular text – the *Letter to King Ptolemy*, communicating the solution to a mathematical problem – crafted as a letter? As has already been noted, letters were written for a wide variety of purposes in the Greco-Roman world. Letter-writing was taught to those boys who were educated. Evidence suggests that in the Hellenistic period, at least at certain social levels, letter-writing had become fairly commonplace (Harris 1989: 127–128; see also Ceccarelli 2013).

Brief letters also served the purposes of mathematicians; such letters were sometimes used as a preface to a technical mathematical text (presented in the characteristic format for such texts; see Taub 2013); a number of such letters attributed to ancient Greek mathematicians survive. In a letter to Dositheus, Archimedes referred to their correspondence: 'Earlier, I have sent you some of what we had already investigated then, writing it with a proof' (ed. Heiberg 1910: vol. 1, 2; trans. Netz 2004: 31). This brief comment serves as an introduction to the text, most of which is presented in the form of problems, propositions and proofs. Archimedes had spent time in Alexandria, and continued his connections there by corresponding with a number of individuals interested in mathematics, including Conon and Eratosthenes, as well as Dositheus. He apparently often sent out enunciations without proofs, that is, puzzles in advance of the works themselves (Netz 2004: 13–14); one such example is the *Cattle Problem*, set in an epigram, as we saw in Chapter 1, and conveyed in a letter addressed to Eratosthenes.

Epistolary Styles

Letters served a wide variety of purposes in antiquity, and had various shapes, from the simple and straightforward to the more elaborated (as is the case with the *Letter to King Ptolemy*; see now Ceccarelli 2013). In some instances, there is a degree of slippage between the idea of a letter and the idea of a preface addressed to an individual. Just as Archimedes' letters could serve as an introduction to a mathematical text, some prefaces appear to be a kind of letter; Pliny the Elder's preface to his *Natural History* is addressed to Titus, and may be read almost as a free-standing epistolary text.

That well-written letters conformed to a particular style is indicated by the comments on style found in writings of Cicero (106–43 BCE), for example, his letter written to Gaius Scribonius Curio (in 53 BCE), where he refers to two styles of letter which have great charm for him: one intimate and humorous, the other austere and serious (*Letters to Friends* (*Epistulae ad Familiares*) 2.4.1 [trans. Bailey 2001: 234–237]). That some people paid attention to the crafting of letters is clear; others were encouraged to do so. Gregory of Nazianzus (fourth century CE) produced a 'style manual', listing some of the features of a well-composed letter. He concluded his remarks (conveyed in a letter addressed to Nicobulus, *Epistle* 51, 7) by noting that there is an 'unadorned quality, which is as close to nature as possible, that must especially be preserved in letters' (trans. Malherbe 1988: 61). But he also recommends the discreet use of embellishment, including proverbs and other devices (*Epistle* 51, 5–6), elements that are found in abundance in the *Letter to King Ptolemy*. From this perspective, the opening poetic quotation of the Delian legend seems particularly apt; in his own time, Eratosthenes was highly regarded as a well-rounded scholar, with strong literary as well as scientific interests. That he would write a letter to the king that would display several different literary styles and modes of embellishment should not surprise us. Part of Wilamowitz' suspicion of the authenticity of the *Letter* was based on his own notions of style and literary convention; he argued that the text was lacking in overall style.[22] But, some aspects of the *Letter* correspond closely to letters authored by Archimedes, himself one of Eratosthenes' correspondents. The *Letter to King Ptolemy* shares several characteristic features of Archimedes' letters, including a simplicity of address and, in the proof presented, technical language and references to diagrams.

Many ancient Greek mathematical texts have a specific character, which will be familiar particularly to students of geometry, exhibiting a particular linguistic style and format of presentation. The use of technical, formulaic language and lettered diagrams may be regarded as key features of such texts (see Netz 1999a), and these features are also present in the *Letter to King Ptolemy*: technical terms – such as 'parallelograms' – are used, mathematical objects are labelled with letters, and there is an implied reference to a diagram. The geometrical demonstration in the *Letter* (ed. Heiberg 1915; trans. Thomas 1939–1941; trans. Drabkin, in Cohen and Drabkin 1948) shares the same basic format, with some variation, familiar from other geometrical texts. Overall, the proof outlined in the *Letter* employs the conventions expected in a mathematical text.[23]

---

[22] Wilamowitz-Moellendorff 1894/1971: 53: 'Dieses Machwerk hat überhaupt keinen Stil'.
[23] For a more detailed discussion of the proof given in the *Letter*, see Taub 2008b.

But there are other facets of the *Letter* that are more literary in character. Indeed, Knorr suggested that the text 'is of greater literary than technical interest'. Several scholars, including Carl Huffman, have noted that the Delian problem 'bears some of the signs of a literary composition'. Whether or not the legend was based on fact is not an issue here; that it makes a good story, and is open to divergent interpretation, contributes to the literary charm and, possibly, ambition of the text (Knorr 1989: 131; Huffman 2005: 377; Wilamowitz- Moellendorff 1894/1971: 48 thought the story plausible).

Furthermore, the literary appeal of the work as a whole is amplified by the inclusion of two pieces of poetry, strategically placed at the beginning and the end: the quotation near the opening, from an ancient tragedian, unnamed but identified as such (even though the text cannot be traced to a known author); and the closing epigram, which serves as an elegant conclusion to the *Letter*:

> If you plan, of a small cube, its double to fashion,
> Or – good sir – any solid to change to another
> In nature: it's yours. You can measure, as well:
> Be it byre, or corn-pit, or the space of a deep,
> Hollow well. As they run to converge, in between
> The two rulers – seize the means by their boundary-ends.
> Do not seek the impractical works of Archytas'
> Cylinders; nor the three conic-cutting Menaechmics;
> And not even that shape which is curved in the lines
> That Divine Eudoxus constructed.
> By these tablets, indeed, you may easily fashion –
> With a small base to start with – even thousands of means.
> O Ptolemy, happy! Father, as youthful as son:
> You have bestowed all that is dear to the Muses
> And to kings. In the future – O Zeus! – may *you* give him,
> From your hand, this, as well: a sceptre.
> May it all come to pass. And may he, who looks, say:
> 'Eratosthenes, of Cyrene, set up this dedication'.[24]

The quotation from the tragic poet and the epigram together confer a special degree of literary interest and distinction to the *Letter*. Indeed, modern critics have been much impressed by the epigram: Rudolf Pfeiffer referred to it as 'faultless and gracious', as well as 'formally perfect', noting that the mathematician was recognisable in this formal perfection (Pfeiffer

---

[24] Trans. Netz 2002: 214 (and also 2009a: 163). As noted earlier, historians generally agree that the quotation that purports to be from the monument is genuinely the work of Eratosthenes.

1968: 156; 168). And the closing lines, naming Eratosthenes of Cyrene, provide a personalised 'seal' (*sphragis*) to the letter as a whole.

Signing and Sealing the *Letter*

The sealing of letters and other documents using wax or lead with a stone or metal seal was an important practice in ancient Greece and Rome, being employed in a manner similar to the modern use of the signature on documents (Pryce, Strong and Vickers 2012). The literary use of a 'seal' (*sphragis*) as a closing to poetry is well attested as a way of personalising a text (as with a signature), analogous to the use of a wax seal on a piece of papyrus (Thesleff 1949; see also Aly 1929 and Kranz 1961).[25]

Examples of poetic 'seals' can be found in a number of early Greek poems, including Hesiod's *Theogony*; Deborah Roberts has suggested that the early function of the poetic *sphragis* 'may in part have been to identify a work or body of work with its author in a period of poetic fluidity' (Roberts 2012). Holger Thesleff pointed to the Homeric Hymn to the Delian Apollo, in which 'an invocation with mythological stuff is here followed by an ample personal passage'; the poet (lines 166–173) identifies himself to the inhabitants of Delos, particularly the chorus of maidens:

> Think of me in future, if ever some long-suffering stranger comes here and asks, "O Maidens, which is your favorite singer who visits here, and who do you enjoy most?" Then you must all answer with one voice(?), "It is a blind man, and he lives in rocky Chios; all of his songs remain supreme afterwards."[26]

The identity and status of the poet are emphasised through the 'seal'; Thesleff explained that 'a condition of sphragis of the Delic hymn is the fact that the poet's self-consciousness rose when his public changed. The anonymous epics were delivered in the presence of nobles, but hymns like *nomos* [a particular style of song] in the presence of the people, among whom the poet could be an important person' (Thesleff 1949: 120–121).

While several early extant Greek poems were closed with a poetic *sphragis*, the persistence of certain literary forms allowed later poets to adopt a similar practice and to develop the use of such a 'seal'.[27] Thesleff notes that

---

[25] Fraser 1972: vol. 1, 412 n. 289 (text of the note is in vol. 2a, 594 (note 289)) mentions the 'seal' in relation to Eratosthenes' epigram, and cites other relevant works.
[26] Trans. West 2003: 85.
[27] See, for example, Bacchylides *Ode* 3 (trans. Campbell 1992: vol. 4, 135), and its closing reference to Ceos, the birthplace of Bacchylides (*fl.* fifth century BCE).

'the author sometimes brought matters of his own up to discussion when he had the opportunity of introducing himself'; this is just the sort of thing we see in Eratosthenes' epigram. Thesleff (1949: 128) suggested that the 'sphragis is a sort of advertisement'; the example in Eratosthenes' epigram strongly supports this appraisal. The *sphragis* was not simply part of the oral poetry tradition, but was adapted to a culture of writing and reading. While audiences would have heard the earlier poets recite their work, the *Letter to King Ptolemy*, with the epigram, would have been read, possibly out loud; the monument described in the epigram would have been encountered by those going past, who may have stopped to look at the inscription. The *sphragis* allows the identification of the poet; it is a convention of many ancient Greek mathematical texts to use the first-person pronoun when describing the solution of mathematical problems, just as the author of the *Letter to King Ptolemy* has done. The *Letter* incorporated the conventions of both of these types of texts in crediting the 'author' with his creative work, both mathematical and poetical.

The use of a *sphragis* to close the epigram and the entire letter may also reflect both Eratosthenes' interest in Homer, and his use of 'seals' in his geographical work, and also – through these associated interests – serve to further personalise the epigram and the letter. According to Strabo (2.1.22), Eratosthenes used 'seals' in the *Geographica*, but in a different way from the poetic use: he employed 'seals' (*sphragides*) as geometrically defined shapes to map geographical space; India was described as 'rhomboidal' (15.1.11).[28] In the modern period, Eratosthenes is most renowned for his geographical work, but he considered his literary-critical work to be important as well.[29] His *Geographica* does not survive, but we know from Strabo (1.2.3) that the work opened with a consideration of the Homeric poems, and their relevance for geographical knowledge (see Fraser 2012 and Pfeiffer 1968: 163–168, esp. 166, on Eratosthenes and 'Homeric geography'; see also Roller 2010). According to Strabo (1.1.10; 1.2.3), it is in the context of his consideration of Homer and the geographical information

---

[28] Strabo (2.1.30) explains that it is possible to describe a country's shape with reference to a geometrical figure (such as a triangle), or a well-known figure, like an ox-hide or the leaf of a plane-tree. Geus 2004: 25 argues that Eratosthenes had a didactic purpose and aimed to produce a map that was easy to memorise and copy.

[29] Suetonius notes that Eratosthenes was the first to use the title 'Philologos' for himself; cited by Pfeiffer 1968: 158, n. 8; see Suetonius *De grammaticis et rhetoribus* 10, ed. Brugnoli 1963: part 1, 14–15. See also Pfeiffer 1968: 156. In addition to his (probably) twelve-book work on comedy, he wrote a number of poems; Plutarch *Quaestionum convivialium* VII 1, 3 (699) called him 'kompsos', meaning 'elegant' and 'cultivated' (see Geus 2002: 99).

he offers that Eratosthenes offered his own view that the aim of the poet is not to instruct but to give pleasure.

To some extent the epigram that closes the *Letter* serves to turn the authority of the past on its head: this is not only an historical or 'popular' account, but an announcement and celebration of a discovery, an innovation. From this perspective, the *Letter* can be understood as a heuristic text, announcing another 'Eureka' achievement. The epigram's *sphragis* – alluding simultaneously both to Homer and to Eratosthenes' own *Geographica* – is a very suitable closing, seal or signature, to the entire text of the letter, announcing his discovery of a solution to the Delian problem. And both the letter's salutation and the closing lines of the text – the signature – link Eratosthenes directly with Ptolemy, his royal patron.

Why Send a Letter If You've Built a Monument?

While acknowledging that the *Letter to King Ptolemy* is a rich and complicated mathematical text, it still seems important to us to confront the question as to why the author chose to communicate his solution of a geometrical problem in this way – in a letter. Michael Trapp has noted that it was not unusual for Greek technical and scholarly works to be written in epistolary form, on various subjects, including mathematics, medicine and philology; he cites, as an example, the letters of Dionysius of Halicarnassus to Ammaeus and Gnaeus Pompeius on rhetoric and writing style (Trapp 2012). But the *Letter to King Ptolemy* is unusually complicated in comparison to many of the other letters that survive on mathematical topics, particularly with its inclusion of the historical account, the quotation of poetry, and the description of the votive monument.

In fact, there is some irony in the fact that the letter, often thought to be a rather ephemeral form of communication, has been preserved while the expensive monument it describes does not, apparently, survive (if in fact it ever existed). Furthermore, in antiquity, as now, the letter was often a public form of communication, in a similar way to how one might publish an 'open' letter, or send a 'letter to the editor'. This *Letter* was probably not a strictly private communication. What might have motivated Eratosthenes to write it?

To return to the features of the letter as form of communication, the *Letter to King Ptolemy*, if it was actually delivered, would have been written, probably dictated, by the author (presumably Eratosthenes) on papyrus, which would then have been rolled up and sealed. As Librarian in

Alexandria, and royal tutor, Eratosthenes could have called on the services of a courier to deliver the letter. One of the principal aims of the *Letter* is to describe the monument that Eratosthenes erected to celebrate his discovery and to dedicate this to his royal patron. P.M. Fraser has noted that 'the monument itself reflects a not uncommon habit in antiquity of executing advertisements of scientific achievements, in the form of such dedications'.[30] Wilamowitz argued that, if Ptolemy had the monument, he would have had no need for the letter, but he regarded the inscription attributed to Eratosthenes as authentic (Wilamowitz-Moellendorff 1894/1971: 52; cf. Fraser 1972: vol. 1, 412). However, Huffman noted that it is not clear that the monument would have been located in a place where the inscription would be easily legible; the letter provided the king with a conveniently readable version. Furthermore, the inclusion of the history of the Delian problem in the *Letter* would have made the significance of Eratosthenes' achievement clearer to the king, as well as to anyone else who encountered the complete text. For it is unlikely that the *Letter* was intended only for the king; it may have served as a method of publication of Eratosthenes' discovery. The monument could hardly have been reproduced and disseminated to others, while the *Letter* could be, as it eventually was in Eutocius' anthology of solutions to the problem (Huffman 2005: 373).[31] Here, what may have originally been intended as an 'occasional' piece was valued sufficiently to have been preserved, eventually to be brought together with other solutions, and anthologised (cf. Rosenmeyer 2006: 25). (Remember, also, that epigrams themselves were often composed as 'occasional' pieces; the existence of the *Greek Anthology* is testimony to a powerful inclination to collect and preserve for posterity.) In the case of the *Letter to King Ptolemy*, it seems likely that the author – whoever he was – was also writing for future ages.

The *Letter to King Ptolemy* celebrates Eratosthenes' solution of the problem of the duplication of the cube, but the function of the *Letter* was not simply to share the proof. By examining the text as a whole, we see that the epistolary format allows the author to incorporate a multiplicity of other genres within the text and, in so doing, to draw out links to a number of

---

[30] Other monumental inscriptions erected by ancient mathematicians and philosophers include the Canobic inscription attributed to Claudius Ptolemy and that erected by Diogenes of Oinoanda setting out Epicurean doctrines (Fraser 1972: vol. 1, 412–413). On the Canobic inscription, see Hamilton, Swerdlow and Toomer 1987; on Diogenes of Oinoanda, see Clay 1990 and M.F. Smith 1993; 1996; 2003.

[31] Fraser 1972: vol. 1, 413 considers where the monument may have been placed.

intellectual traditions, including the mythographical and geographical, as well as the mathematical and the literary.[32] The *Letter to King Ptolemy* celebrates Eratosthenes' solution to a famous mathematical problem; it also points to his achievements in other areas of intellectual pursuit, notably textual and historical studies, as well as geography and poetry. The *Letter to King Ptolemy* is not a simple document, nor is it simply a letter. It richly incorporates several subtexts presented in a mixture of modes, and testifies and pays tribute to Ptolemy's good judgment in patronising Eratosthenes' intellectual versatility. The *Letter* demonstrates the potential complexity of the epistolary format, and suggests that, in the Hellenistic world, mathematics was of interest not only to mathematicians; in this way, the *Letter* serves as an elaborate 'envelope', presenting a collection – or special package – of material of interest not only to historians of mathematics.

Conclusion

The letter was an important genre of text in Greco-Roman antiquity, used for various purposes. The natural-philosophical and mathematical letters discussed here all served to convey not only information but also some sort of gift, be it a philosophical epitome to aid in gaining tranquillity, a tantalising gift of a mathematical problem or puzzle, or an elaborately constructed many-layered confection, the *Letter to King Ptolemy*. An important feature of the genre of letter is the deliberate act of sharing, whether with an individual (one's friend, follower or patron) or a community (fellow Christians; fellow mathematicians). Scientific and mathematical letters imply the existence of a community of shared interests and expertise; the intertextuality found in these letter-texts reinforces the ties between members of communities, who share these and other texts.

As a type of text, the letter enables an author to emphasise and reinforce relationships to others, be they patrons, followers, colleagues or rivals. By emphasising relationships to the addressee, the author signals engagement with others, sometimes common cause, sometimes competition. And as both Cicero (*Letters to Atticus* 8.14.1; 9.10.1; 12.53) and Seneca (*Moral Epistles* 75, 1) pointed out, letters are sometimes like a conversation, or a portion of such, once again pointing to multiple tensions between oral discourse and written texts.

As was noted earlier, the genre of the letter provided space for literary forgeries of various types, especially those attributed to famous figures,

---

[32] See Netz 2009a: 160–164 on the tensions between myth and mathematics in Eratosthenes' text.

including philosophers and mathematicians, such as Plato, Aristotle, Archimedes and Eratosthenes. Yet such literary forgeries are not to be completely shunned, for they can yield valuable historical information regarding the place of scientific ideas and mathematical problems within wider culture; such literary forgeries – whether fictional poses or scientific fantasies – augment our understanding of Greco-Roman intellectual life.

Epilogue

The letter referred to at the opening of this chapter and attributed to Dionysodorus has special resonances with some of the other Greek letters considered here, namely the letters of Epicurus (whose school, the Garden, was known for being open to women as well as slaves) and those of mathematicians, especially that which is ascribed to Eratosthenes. Pliny reports that Dionysodorus was a well-known geometrician, as were some of the authors of our other mathematical letters. Dionysodorus deposited his letter in a tomb, which might remind us of Minos' efforts to create a beautiful mausoleum for his son Glaucus, referred to in Eratosthenes' account of the Delian problem, also presented in a letter. In Pliny's account of the story, the first readers of Dionysodorus' letter were women. While the literacy rate of women in antiquity was not high, images surviving from Pompeii depicting women with writing tablets and pens indicate that in Pliny's world female literacy was valued (Harris 1989: 263).

3

Encyclopaedia

> First and foremost I must deal with subjects that are part of what the Greeks term an 'all-round education' [*enkuklios paideia*], but which are unknown or have been rendered obscure by scholarship.
> Pliny the Elder, *Natural History*, Preface 14, trans. Healy 1991: 4.

Gaius Plinius Secundus (known to us as Pliny the Elder, who famously died in 79 CE) was the author of an unusual work, the thirty-seven-book *Historia Naturalis* or *Natural History* (*NH*), which holds a significant place in several fields of intellectual history, including history of science. The *Natural History* is the only one of Pliny's works that survives; it is often referred to as an 'encyclopaedia'. In considering this work, and its status as a new 'genre', I first ask what it means to be an 'encyclopaedia'. Second, I consider recent work by others pointing to links between Pliny's work and Roman imperialism. Finally, I turn to examine Pliny's treatment of some scientific material, specifically meteorology, to explore the ways in which his approach is 'encyclopaedic' and imperial. The *Natural History* is a particularly interesting case, because while other encyclopaedic writings may have been produced in antiquity (depending on how we define 'encyclopaedia'), the *Natural History* is the only one to have survived in its entirety. The *Natural History* stands at the beginning of a tradition that developed and flourished primarily in later periods.[1]

To turn to my first question: What does it mean to be an 'encyclopaedia'? Here, I am not attempting to develop and offer a theory of genre in general, or even of genres specific to texts which communicate scientific and mathematical ideas. As mentioned in the Introduction, a useful working definition of genre is 'a recurring type or category of text, as defined by structural, thematic and/or functional criteria' (Duff 2000: xiii). It seems reasonable to begin a consideration of the possible genre referred

---

[1] Chibnall 1975; Nauert 1979, 1980; Reynolds 1983 have studied its influence in later periods.

to as 'encyclopaedia' by looking at form, content and function. But before considering such issues, we should consider the label 'encyclopaedia'. For some, the term evokes the great eighteenth-century French project, the *Encyclopédie* of Denis Diderot and Jean le Rond d'Alembert. Sorcha Carey (2003: 17) notes that Rabelais was the first to use the term 'encyclopaedia', in the sixteenth century. In using the encyclopaedic format, Pliny provided an important model for later work and for the communication of scientific information and ideas. Trevor Murphy (2004: 215) has argued that 'to the Renaissance and subsequent eras, the *Natural History* was a precedent for how the diversity of multiple intellectual traditions and discourses could be reassembled into a universe, under imperial authority, in a single all-receiving text'.

In the Preface to the *Natural History*, Pliny refers to what he says 'the Greeks call *enkuklios paideia*'. In describing his work, Pliny explains that his own

> path is not well-worn by writers, nor the kind along which the mind aspires to wander. No Roman author has attempted the same project, nor has any Greek treated all these matters single-handed ... First and foremost I must deal with subjects that are part of what the Greeks term an 'all-round education' [*enkuklios paideia*], but which are unknown or have been rendered obscure by scholarship.[2]

It is not entirely clear what subjects or studies fall into Pliny's concept of *enkuklios paideia*, nor is it clear that he is referring to his own work in those terms. Some scholars have taken the term to refer to the Greek 'system' of education in core subjects; Aude Doody has addressed the problems in reading the phrase in Pliny, and has suggested that Pliny sets his own work in the context of Greek polymathism.[3]

Petrus Johannes Enk suggested that in Greece 'the Sophists were the first who claimed to impart to pupils all the knowledge they might want in daily life', but that, nevertheless, 'notwithstanding the value the Greeks attached to [what we might term] encyclopaedic knowledge, they never got so far as to compose an encyclopaedia'.[4] Rather, as Enk explained, 'it

---

[2] *NH* Preface 14, trans. Healy 1991: 4. Rackham 1949/1979 1: 11 translates *enkuklios paideia* as 'Encyclic Culture'.

[3] Doody 2010: 51 notes that 'the point of Pliny's invocation of *enkyklios paideia* is to set the *Natural History* in the context of abstruse Greek knowledge'. Morgan 1998 has argued that *enkuklios paideia* was the average education available to all.

[4] Enk 1970: 383. He suggests that when Quintilian (*Institutio Oratoria* 1.10.1) speaks of *enkuklios paideia*, 'he means the ordinary course of instruction for a pupil before taking up his special subject of study'.

was reserved for the practical-minded Romans to lay down in a compilation the results attained by the scientific researches of the Greeks' (Enk 1970: 383). Several Latin authors are often cited as having written encyclopaedic works, including Cato, Varro and Celsus. However, the extent to which any of their writings would qualify as examples of an encyclopaedia is not entirely clear. More recently, scholars have argued that these works are rather different from Pliny's, in that they divide their subjects into *artes*, such as agriculture, military science, rhetoric and medicine (Doody 2010: 53; Codoñer 1991; Murphy 2004: 13).

Pliny does not seem to be offering a system of 'general education' in the *Natural History* (Doody 2009: esp. 12, 16). Enk described Pliny's encyclopaedia as being a work on nature *and* visual art and here hinted at a distinction between those Roman writers whose works discuss the various *artes* and Pliny. In the first group, the works by Cato, Varro and Celsus (such as they were) might be included, while the second category seems to have had only one example which dealt with the physical world: Pliny's *Natural History* (cf. Doody 2010: 55–58). Varro, Cato and Celsus may have produced works inspired by the Greek educational aspiration of *enkuklios paideia*, but that is not the same as having produced an encyclopaedic work, or an encyclopaedia.

Returning to issues relating to the notion of 'genre' as understood by modern literary theorists, there are questions as to how to understand the *Natural History* as an 'encyclopaedia'. Turning to characteristics of theme, structure, and function, we find that one of the principal themes of the work, the subject that it is meant to cover, is famously announced by Pliny in his Preface: 'Nature, that is, life, is my subject' (*rerum natura, hoc est vita, narratur*, Preface 13). Mary Beagon has explained that for Pliny *natura* meant 'the world, both as a whole and as its separate components'; *natura* is everything (Beagon 1992: 26). Trevor Murphy has suggested that 'we might just as well translate *Naturalis Historia* as *Inquiry into Everything*' (Murphy 2004: 33). Keeping questions of genre in mind, the title, as Murphy recommends we understand it, is not much different from Lucretius' poem *de rerum natura*, yet the two works are markedly different in terms of formal structure. In other words, the title, or even the subject matter, is not likely to be terribly helpful in understanding the genre of Pliny's work.

The question of the structure and organisation of the *Natural History* is somewhat problematic; Doody has studied the structure of the work as a whole, and particularly examined the role of the *summarium* as an aid to finding information contained within the text. (The *summarium* is found in Book 1, and lists the contents of the other thirty-six books, along with the sources consulted by Pliny.) Doody emphasises that Pliny did not intend for readers to treat the text as a simple, straightforward narrative; in fact, Pliny did not expect readers to read the whole work (Doody 2010: 92). Some scholars, including Murphy, have acknowledged that 'the book does have organizing principles ... laid out in great detail in book 1', but he argued that the 'train of thought is often interrupted, since Pliny is usually willing to be diverted from the topic in hand in another direction by association of ideas'; on Murphy's reading, the structure of the *Natural History* is digressive (Murphy 2004: 30–32). But Doody has argued that 'there is a logic to the *Natural History* that only reveals itself to the reader who follows the stream of information from fact to fact, section to section, book to book, subject to subject' (Doody 2010: 92). Doody's understanding of the structure of the *Natural History* relies on comprehending the work in terms of its genre, that is, comprehending the work as an encyclopaedia.

The theme of the *Natural History*, 'all of nature' or 'everything', and, to some readers, the seemingly digressive structure probably do not offer sufficient information to qualify the work as representing a separate genre. Rather, it is the *function* of the work that provides a key to understanding what an encyclopaedic work may have signified to Pliny. The *Natural History* was 'meant not just for reading, but for a particular kind of reading, that is, reference or reading for use'; 'it anticipates being consulted on points of detail (pref. 33)'.[5] Its function and usefulness as a detailed source of reference are the keys to understanding the genre of 'encyclopaedia'. Pliny, in the *Natural History*, has imposed a structure, an order, on the world, on nature, and provided a special tool, the *summarium*, which can be used to refer to and extract the useful information collected in the work.

That the *Natural History* was an important and quintessentially Roman work is agreed by most scholars. During the 1990s, several key studies, including those by Mary Beagon, Roger French, and Andrew Wallace-Hadrill, highlighted the themes of *natura* and *Romanitas* as present in Pliny's work. Recently, several scholars have considered the imperial context and aims of the work. Sorcha Carey examined Pliny's geography as a preamble to concentrating on his history of art and its relationship to Roman culture; she emphasises Pliny's eagerness to assert the 'romanness'

---

[5] Murphy 2004: 12 and 196; see also 211, where he comments that the *Natural History* is not a didactic work, even though it contains detailed information.

and 'imperialness' of the world he describes. In her view, Pliny's text can be regarded as 'the greatest expression of triumphalism', for here Pliny – that is, implicitly, Rome – 'has collected the entire contents of the world'. The *Natural History* aims to represent the conquering and synthesising of the entire world, to emphasise the domination of the world by Rome. As Carey puts it: 'knowing the world ... is tantamount to conquering it'. The genre of the encyclopaedia is particularly well suited to establishing a conceptual link between knowledge of the world and conquest; in her view, the encyclopaedia 'is a genre in which not only knowledge itself, but the instruments of knowledge – lists, taxonomies and sources – reign supreme'. Pliny's catalogue of the world is a catalogue of the Roman empire (Carey 2003: 33, 76; 39–40).

Murphy also regards the encyclopaedia as a Roman development linked to empire. Focusing on the *Natural History*, he argues that 'to gather this large and miscellaneous body of knowledge into one book and subordinate it to a central set of principles' was the innovation of the Romans. His explanation of the genre emphasises the role of conquest and incorporation:

> *Encyclios paideia* was by origin alien to [the Romans]. To the Classical and Hellenistic Greeks, it was to be taken for granted as part of the matrix of their native culture; to the Romans, it was an intellectual component of the spoils of war. The digestion and assimilation of *encyclios paideia* to the logic of Roman society could only be executed by a deliberate and self-conscious effort of transposition.

Acknowledging that the Roman tradition of encyclopaedism added practical lore, such as that useful for agriculture, to Greek subjects like rhetoric and philosophy, Murphy argued that 'creating the first real encyclopaedia required the command of vast intellectual territories, and, as the history of the times when it was written suggests, required also the ambitions and the far-reaching mental horizons attendant on administering an empire' (Murphy 2004: 195). In fact, Pliny himself enjoyed a diverse career;[6] at the time of his death he was commander of the imperial fleet at Misenum, and was personally familiar with administrative issues.

But there remains a huge question about how to define 'imperial'. There are, certainly, a number of features of the *Natural History* that may serve as literary examples of Roman imperialism, for example the conquest and

---

[6] Pliny's career included high procuratorships and membership of the inner 'cabinet' of Emperors Vespasian and Titus. Flavian dynasty: Vespasian 69–79, Titus 79–81, Domitian 81–96.

synthesis pointed to by Carey and Murphy, as well as the addressing of the Preface to the Emperor himself (Titus 79–81). Furthermore, the emphases on the imposition of order and the extraction of resources – both understood as vital to the Roman peace – were fundamental aims of the empire, and are both crucial to Pliny's work as well. To be organised in an orderly fashion and to encourage the extraction of what is useful: these are the hallmarks of Pliny's presentation and use of the genre of the encyclopaedia.

But what can we say about the encyclopaedia as a genre used for communicating scientific ideas and methods? Pliny was not an expert in any particular area, nor was he a natural philosopher. Rather, he has often been understood as a self-conscious populariser, providing information to a wide audience composed of educated men. Many scholars have disparaged Pliny as a 'mere' compiler, but Pliny presents a summary not only of contemporary knowledge, but also of different points of view. And, in at least one area of ancient scientific thought, that is, meteorology, Pliny is original, precisely because he is 'encyclopaedic'. In the case of the *Natural History*, 'encyclopaedic' means Roman and imperial, that is, concerned with usefulness and order and the ability to access and extract resources; by 'encyclopaedic' I also incorporate notions of expansion, assimilation and synthesis.

Pliny seems to be very comfortable with collecting, presenting and appropriating the scientific 'results' of Greek predecessors (including data and theories), holding the view that such results should be applied by Romans to Roman concerns. Historians of science have often disparaged the scientific content or value of Pliny's work, but we need to remember that, above all, Pliny's project is a Roman enterprise, conducted within the context of the Empire; his project differed in fundamental ways from many of those initiated by Greek authors. Even though the label 'encyclopaedia' seems to hark back to some Greek notion, as Enk, Doody and Murphy have argued, the encyclopaedia may be understood as a Roman, even a specifically imperial, genre. Pliny's development of a new 'genre' of learned communication – at once comprehensive, imperial and entertaining – should alert us to functional, as well as possibly structural and thematic, differences with other ancient genres of scientific communication.

Pliny (Preface 13–14) claims to have gone where no Roman or Greek had gone before; he asserts that his 'path is not a beaten highway of authorship', suggesting that he believed he was doing something really different from his predecessors. His claim to have gone where no one else had previously ventured does indeed seem to be justified by the *Natural History* as a whole, as an important example of a new genre, the encyclopaedia. And, when we look specifically at his treatment of meteorology, we discover that what Pliny was doing was radically different from other ancient authors writing on the subject. Ancient Greco-Roman meteorology was defined by two seemingly parallel traditions: one concerned primarily with prediction, the other with explanation (Taub 2003). Exceptionally amongst Greek and Roman authors, Pliny was concerned *both* with prediction and with explanation. Here he is certainly justified in saying 'there is not one person to be found among us who has made the same venture, nor yet one among the Greeks who has tackled single-handed all departments of the subject' (trans. Rackham 1: 9–11).

Works in each of these separate Greco-Roman traditions of meteorological prediction and explanation tended to emphasise that what was being presented *was* indeed part of a tradition, in many cases extending back centuries. In fact, the reliance on the work of predecessors, both in shaping predictions and explanations, is a key characteristic of Greek and Roman meteorology. The efforts of many people were incorporated into the ancient texts, which thus convey a sense of community, in which authors incorporate and build upon the work, including the ideas, observations and prognostications, of previous contributors to the shared project of predicting and explaining weather. The ancient Greek and Roman texts show the individual authors and their predecessors working together to come to terms with meteorological phenomena (Taub 2003).

In many ways, Pliny's encyclopaedic project represents both the convergence and culmination of Greek and Roman traditions of predicting and explaining meteorological phenomena.[7] As he informed us, in the course of compiling and organising the *Natural History*, Pliny relied heavily on the work of others; his *summarium* proudly proclaims his use of these sources. He claimed (Preface 17) to have included more than 20,000 facts culled from more than 2,000 works. To an important extent Pliny can be understood as having brought together (that is, ordered) and made available (allowing extraction and use of) what had already been collectively accomplished in the fields of weather prediction and meteorological explanation.

G.E.R. Lloyd, perhaps somewhat disparagingly, described the *Natural History* as being primarily based on reading, rather than observation (Lloyd 1983: 135–149). We will return to this point. For the moment it is sufficient to note that with reference to meteorology, the links within the

---

[7] See Murphy 2004: 15 on Pliny's bringing together of new and traditional knowledge more generally.

Natural History to other authors are crucial, but there is also an underlying tension present. Pliny, while working to a great extent within these literature-based traditions, shares his own sense of what he regards as the proper Roman approach to meteorology, which is to some extent to ignore the literature, the books and the (largely Greek) authors, and to argue for relying on (good old) Roman skills, such as observation.

As already noted, in his dual aims of explaining and predicting meteorological phenomena, Pliny is unusual. Given the organisation of the *Natural History*, and the way it was compiled, it is not surprising that his discussions of the prediction and explanation of meteorological phenomena are largely presented in separate sections of the *Natural History*, as his own sources would have treated meteorological phenomena *either* from the point of view of explanation *or* from that of prediction; Pliny treats both prediction (as part of agricultural skill in Book 18) and explanation (part of natural philosophy, in Book 2). Pliny was not particularly philosophical in his approach. Rather, he was interested in practical issues. Furthermore, Pliny incorporates material not only from Greeks and Romans; he cites the views, for example, of Babylonians and Egyptians (*NH* 2.79.188; 2.81.191).[8]

Nevertheless, he was discriminating in what he chose to adopt from others; he does not simply present a list of what has been said or suggested by them.[9] While his admiration of Greeks' contributions to meteorology is clear, he does not hesitate to reject some of their ideas.[10] And while he celebrates the observations and curiosity of the Romans, he disparages the lack of work done by his fellow Romans in the field of meteorology, making it clear this should be undertaken as an imperial endeavour:

> More than twenty Greek writers of old have published their observations on these topics. And this surprises me, that although the world was in a state of upheaval and divided piecemeal into kingdoms, so many men were concerned with facts that were so difficult to research – especially when living in the midst of wars, when hosts were untrustworthy and pirates, the universal scourge, held up the free passage of information. Consequently, today a person may learn some facts about his own region, from the notebooks of those who have never been there, more accurately than from the knowledge of the local inhabitants. Yet, nowadays, in this happy time of peace under an emperor who takes such pleasure in promoting literature

---

[8] I tend to agree with Murphy 2004: 196–197 that Pliny's work, unlike that of Varro and Celsus, is not concerned with teachable skills, but rather is a large-scale enquiry into nature.
[9] See, e.g., his discussion of the causes of earthquakes, *NH* 2.81.191-2.86.200; Taub 2003: 185.
[10] For example, at *NH* 2.59.149, he rejects the ideas of Anaxagoras regarding the possibility of divination and stones falling from the sky; see also Taub 2003: 182.

and science, absolutely nothing is being added to the sum of knowledge as a result of original research; indeed not even the discoveries made by people long ago are thoroughly assimilated. (*NH* 2.45.117, trans. Healy 1991: 24)

Pliny's own work on meteorology is intended, to some extent, to rectify the situation.

His account of the causes of meteorological phenomena occupies relatively brief sections, but the discussions are importantly located, in the middle (conceptually) of his wider account of his understanding of the nature and workings of the world. He describes and discusses the causes of what he refers to as the phenomena of the air in Book 2, as part of his larger examination of the world.

In his meteorological explanations, Pliny discusses the observations and ideas of predecessors. For example, Aristotle (one of the sources listed in the *summarium*) is mentioned approvingly as an authority several times in Book 2. Pliny's explanations of meteorological phenomena show the influence both of Aristotle and of his pupil and colleague Theophrastus (who, however, is not named as a source). So, for example, exhalations (central to Aristotle's meteorology) play a role in causing weather, as do the motions of the celestial bodies. Furthermore, following Aristotle, Pliny acknowledges that the causes of meteorological phenomena are not entirely understood, but, like Theophrastus (and in contrast with Aristotle), he finds multiple explanations acceptable. But Pliny was no philosopher, and it is not always clear whether his approval of multiple explanations is granted for philosophical reasons, or whether the offering of multiple explanations is, as it were, part of the overall character of the *Natural History*, that is, providing a range of information and points of view.

Weather prediction is discussed within the context of farming activities in Book 18.[11] As a member of the privileged equestrian order, Pliny, like many members of his class, prided himself on his ability and acumen regarding practical matters. He considered himself to be a man of the earth, and he was genuinely interested in agriculture. In his view, agriculture was the most fundamental human activity (*NH* Book 18).

For Pliny, farming was, to some extent, reliant on astronomy, because of the usefulness of astronomy in determining the seasons, establishing an agricultural calendar, and predicting weather. This application of astronomy to practical matters (particularly agriculture, but also trade and navigation) is key to understanding Pliny's interest in astronomy. Pliny was working within a long-standing tradition, going back to Hesiod,

---

[11] Agriculture, arboriculture and gardening are treated in Books 17–19.

linking astronomy to agriculture for practical benefits, including weather prediction. He explains that 'it is an arduous and a vast aspiration – to succeed in introducing the divine science of the heavens to the ignorance of the rustic, but it must be attempted, owing to the vast benefit it confers on life' (*NH* 18.56.206; trans. Rackham 5: 319). This sentiment conveys his sense of duty, to serve his fellow Romans; the section of the *Natural History* that deals with agriculture is perhaps the only portion of the work that could serve as a textbook (see also Murphy 2004: 211).

A key section in Book 18 is the presentation and discussion of the 'farmer's almanac'. Pliny was not content simply to report and reproduce a farmer's calendar that he has found elsewhere. Before even beginning to elaborate his own almanac, he forces the reader to 'submit to contemplation of the difficulties of astronomy, which even experts have been conscious of'. Here, there may be an underlying, implicit tension between the necessity of utilising largely Greek-developed astronomy for the purpose of Roman agriculture; at a number of points Pliny displays caution – even scepticism – about the work of astronomers. He warns his readers that the study of astronomy is not necessarily straightforward and he confronts some of the difficulties involved, outlining, for example, the problems in defining the solar year in relation to the sidereal year, noting that 'it is almost impossible to explain the system of the actual days of the year and that of the movement of the sun, because to the 365 days an intercalary year adds a quarter of a day and of a night, and consequently definite periods of the stars cannot be stated'. He carries on this discussion at some length, indicating that he has given it some thought and that the matter has caused him concern (*NH* 18.56.201-223; esp. 206–207; trans. Rackham 5: 317–321).

Pliny's own farmer's calendar is lengthy and detailed, and includes star phases and associated weather predictions, as well as agricultural advice. He credits various Roman authorities, including Virgil, Cato (234–149 BCE) and Julius Caesar (100–44 BCE) with providing specific information and recommendations.[12] His coverage of the topic begins with the question of the proper date for sowing crops, a question which he states 'needs very careful consideration' (*NH* 18.56.201, trans. Rackham 5.317). Significantly, he notes that 'Hesiod, the leader of mankind in imparting agricultural instruction, gave only one date for sowing, to begin at the

---

[12] The inclusion of Julius Caesar is noteworthy, since he was particularly interested in calendars and (apparently) employed the Greek astronomer Sosigenes to help devise the so-called Julian calendar, named after him. On Sosigenes see Neugebauer 1975: vol. 2, 575.

setting of the Pleiads', but Pliny makes it clear that this date is determined by the location in which Hesiod lived, Greek Boeotia.[13]

Here is a passage from Pliny's own calendar (*NH* 18.65.237, trans. Rackham 5: 339), in which he offers specific dates, stellar risings and settings, and weather predictions:

> Between the period of west wind and the spring equinox, February 16 for Caesar marks three days of changeable weather, as also does February 22 by the appearance of the swallow and on the next day the rising of Arcturus in the evening, and the same on March 5–Caesar noticed that this bad weather took place at the rising of the Crab, but the majority of the authorities put it at the setting of the Vintager.[14]

He goes on to explain that (*NH* 18.65.238, trans. Rackham 5: 341; my emphasis):

> This space of time is an extremely busy period for farmers and specially toilsome, and it is one as to which they are particularly liable to go wrong – the fact being that they are not summoned to their tasks on the day on which the west wind *ought* to blow but on which it actually does begin to blow. This must be watched for with sharp attention, and is a signal possessed by a day in that month that is observable without any deception or doubt whatever, if one gives close attention.

In using the almanac, Pliny cautions against going too closely 'by the book': it is important to make one's own observations.[15] The various weather signs he provides show the sorts of indications to be looked for. The farmer should not be too much in thrall to astronomers and authorities; he should try to use their wisdom, when appropriate, but not be content to stick to 'rules'. Rather, he must look out (literally) for himself. Pliny warns that farmers must take note of the actual day on which the west wind blows, regardless of the calendar date, and attend to appropriate tasks. He emphasises the necessity of careful observation by the farmer himself. Pliny is clear that astronomical knowledge can be useful for agriculture, but emphasises (*NH* 18.75.321) that he is primarily interested in the practical value of general rules rather than in trying to assign particular agricultural operations to specifically stated days. This approach

---

[13] It is worth noting that Pliny names Hesiod, signalling a claim to be operating in the same tradition. Similarly, like Hesiod, Pliny provides practical advice, while being concerned with wider ethical issues. (I have followed Rackham's spelling, instead of 'Pleiades'.)

[14] See Bickerman 1980: 43–50 for an explanation of the naming and numbering of days in any given month.

[15] As noted earlier, Lloyd (1983: 135–149) suggested that the *Natural History* depended primarily on written sources rather than observations.

is reminiscent of the Roman approach to conquered lands more generally: rules (that is, Roman-imposed order) must be put in place, but local conditions also play an important role.

Pliny was well aware of his relation to and reliance on the tradition of the astrometeorological calendars, but he raised questions regarding their usefulness.[16] So, for example, he complains about the difficulties resulting from various authors having made astronomical or meteorological observations at different locations, explaining – at some length – that observations and predictions cannot be generalised from one place to another, noting that the different locations at which individuals worked would contribute to there being differences in what was observed (*NH* 18.57.210–14). He also noted that even astronomers working in the same region (*in eadem regione*) did not always agree with each other.[17] Such a warning would carry special significance for landowners spread throughout the far-flung empire: pay attention to local conditions, and make your own observations.

Pliny makes it clear that, while astronomical knowledge can aid farmers in predicting the weather, they should not be blindly reliant on astronomical expertise. In his presentation and discussion of his own farmer's almanac, Pliny proudly acknowledges his reliance on other types of knowledge. Pliny regards and presents himself as part of an old, largely Roman, tradition rooted in the agricultural life of the land and very deliberately links himself to the esteemed group of ancient authors, including Hesiod, Virgil and Columella, who emphasised the value of agricultural life and who passed on other types of weather lore, such as the behaviours of birds and other animals, in addition to suggestions linked to astronomically based calendars. Although astronomy might be useful to the farmer, astronomical knowledge was not sufficient.

In a strikingly vivid passage Pliny suggests that knowledge of astronomy may not even be necessary for the farmer:

> [Nature] had already formed the remarkable group of the Pleiads in the sky; yet not content with these she has made other stars on the earth, as though crying aloud: 'Why gaze at the heavens, husbandman? Why, rustic, search for the stars? Already the slumber laid on you by the nights in your

---

[16] Pliny is, at times, aware of the problems inherent in relying on the work of others – this in spite of his credulity regarding 'tall tales' and wonders; see also Purcell 2012.

[17] *NH* 18.58.212-13, trans. Rackham 5: 325. He pointed to the following example: 'the morning setting of the Pleiads is given by Hesiod – for there is extant an astronomical work that bears his name also – as taking place at the close of the autumnal equinox, whereas Thales puts it on the 25th day after the equinox, Anaximander on the 30th, Euctemon on the 44th, and Eudoxus on the 48th'.

fatigue is shorter. Lo and behold, I scatter special stars for you among your plants ... I have given you plants that mark the hours, and in order that you may not even have to avert your eyes from the earth to look at the sun, the heliotrope and the lupine revolve keeping time with him. Why then do you still look higher and scan the heavens themselves? Lo! you have Pleiads at your very feet'.[18]

This passage is embedded in a discussion of the generosity of Nature, which has provided many sorts of signs, including insects, stars and plants, to provide information to humans, if only they pay close attention. The study of Nature is the subject of the *Natural History*; all of Nature is to be examined, not only one or another part or aspect. So, a reliance on one type of knowledge becomes automatically suspect.

For Pliny, what was largely Greek astronomy was a useful tool for weather prediction, but could not be relied on by itself. Rather, the ideal Roman farmer included astrometeorology as one of several time-honoured techniques to be used in determining the course of the agricultural year. At 18.224–320 Pliny presents a farmer's calendar, based on Caesar's, as well as other sources. Agricultural advice is included, along with the star phases and weather predictions. At the end of the almanac, Pliny lists signs useful for predicting the weather, including those given by the sun, moon, stars, meteorological events (such as thunder and lightning), fire (including flickering or sparking), the sea, animals and birds. His treatment is, for the most part, limited to providing lists of useful recommendations for landowners and their workers. Pliny was clearly unwilling to rely solely on the advice of 'expert' Greek astronomers; he emphasises the practical necessity of careful observations being made in individual and specific locations. More generally, this emphasis on the importance of local conditions and custom was part of the key to the success of the Roman empire, and the lesson of the empire was not lost on Pliny, for whom specificity and detail are crucial. One cannot always generalise: if one could, the detailed information contained in the *Natural History* would not be useful, or necessary.

The *Natural History* represents the unification of diverse meteorological traditions under a central authority, that is, the Roman Empire. Displaying a network of intellectual and practical relationships (Murphy 2004: 207), Pliny brought together the two different traditions of

---

[18] *NH* 18.67.251-53; trans. Rackham 5: 349. The name 'heliotrope', and the Latinised *heliotropium*, derive from the Greek *heliotropion*: *helios* ('sun') and *tropos* (turning, from *trepein*, 'to turn'); see *OED* online, s.v. 'heliotrope' (www.oed.com/view/Entry/85617?redirectedFrom=heliotrope#eid).

Greco-Roman meteorology, ordering a vast range of information, ideas and techniques, with a deliberate view to making this material accessible, retrievable and usable by readers. To some extent the farmer – a prime user of meteorological information – serves as the exemplar of the ideal Roman for Pliny. The duty of the farmer was to be curious and observant, and to be somewhat 'encyclopaedic' in his study of nature, that is, to gather together the necessary information that would be useful to him, from whatever sources.

The charge that Pliny was a 'mere' compiler of the theories and opinions of others has long been accepted. And a focus on some elements of the apparent structure, rather than the function, of Pliny's *Natural History* may lead us to certain erroneous conclusions. However, a stronger, epistemological claim can be made for Pliny's work, which highlights the function of the new Roman imperial genre: the *Natural History* as encyclopaedia demonstrates the interconnectedness between different bodies of knowledge. As in his presentation of knowledge about weather phenomena, Pliny brings together information about both explanation and prediction; he synthesises ideas and techniques from several traditions. Furthermore, at least in the field of meteorology, he makes an important epistemological claim: specialist 'expert' theory-laden knowledge may be of limited value. And reliance on one type of knowledge, for example astronometeorological knowledge, is not sufficient. Rather, knowledge of animal behaviour and plants must also be included in order to achieve the best practical results.

This inclusiveness and interconnectedness is what distinguishes Pliny's encyclopaedic task: it is the creation of a new genre. Indeed, his new genre requires a new epistemology, which emphasises the interconnectedness of all things within Nature. His genre, the encyclopaedia, is by necessity a compilation, in order to transcend the limits of the individual *artes* and the limiting approaches of his predecessors. Pliny's genre, the encyclopaedia, establishes the possibility of being able to hold in one's hand, to comprehend, different types and sources of knowledge and to see how they relate to each other. In the case of meteorology, Pliny brings together Greek and Roman ideas, along with those of the Babylonians and Egyptians. This synthesis of previously diverse elements within a comprehensive establishment of order is, for Pliny, an achievement itself worthy of the Roman empire, and the Roman peace, itself.

4

Commentary

> In the meetings of the school he [Plotinus] used to have the commentaries read, perhaps of Severus, perhaps of Cronius or Numenius or Gaius or Atticus, and among the Peripatetics of Aspasius, Alexander, Adrastus, and others that were available. But he did not just speak straight out of these books but took a distinctive personal line in his consideration ... He quickly absorbed what was read, and would give the sense of some profound subject of study in a few words and pass on.
>
> Porphyry *On the Life of Plotinus* 14.11–19,
> trans. Armstrong 1966: 41.

The genre of commentary has a long history, extending well beyond the ancient period. For a number of reasons, it has often been treated as a rather desultory sort of text, lacking the excitement of, for example, the dialogue or poem. A number of ancient commentaries on natural philosophical and mathematical texts were composed within pedagogical contexts. For example, Olympiodorus (sixth century CE) produced commentaries on works by Plato and Aristotle as part of his teaching activities in Alexandria. The tendency to regard commentaries as 'school-texts' may contribute to the view that they were not terribly exciting to read. However, not all ancient commentaries were intended for use in formal teaching. Alexander of Aphrodisias, who was appointed public teacher of Aristotelian philosophy probably in Athens sometime between 198 and 209 CE,[1] appears to have had other scholars in mind as his readers. His interest in producing commentaries on Aristotle's works seems to have been primarily philosophical and scholarly;[2] he was known by successive

---

[1] Sharples 2012 notes that this was 'probably though not certainly in Athens'.

[2] Alexander's commentaries on Aristotle's works which do survive include the *Topics*, *Prior Analytics I*, *De sensu* and the *Metaphysics* as well as the *Meteorology*. In contrast, the work of some of the later Neoplatonic commentators appears to have been primarily pedagogical. See Sharples 1990: 95–97; Sharples 1987. Dickey 2007: 49 understands Alexander's commentaries as having been published as written works, rather than as being transcriptions of lectures.

commentators as *the* commentator on Aristotle (Sharples 1994: 1 and 110 n. 2).

The attention given by later generations to earlier writings is key to the work of the commentator. As part of an active scholarly endeavour, commentators engaged critically with their target texts (and their authors) and with other commentators as well. Just as Alexander directed his work towards other philosophers, some commentaries on mathematical and medical works were written by specialists in those fields, presumably for others with similarly specialised interests and competence. Ancient commentators used the genre as a platform for communicating their own new ideas, fresh perspectives and methods; in this way, the commentary was a particularly important genre for communicating natural philosophical, medical and mathematical ideas and practices.

The commentary (often referred to in antiquity as a *hupomnēma*)[3] was one type of text produced within a number of Greek and Roman didactic and scholarly initiatives; epitomes and handbooks are other examples. Commentaries on various texts – including poetry as well as prose works – were especially important from the third century BCE on, and were the products of scholarly activities devoted to understanding texts. As such, commentaries often deal with textual issues, including the explication of vocabulary and terminology, as well as interpretation. As well as those that focused on literary texts, including poetry, the commentary was a particularly significant genre for philosophical (including natural philosophical) writing in later periods. The works of Plato and Aristotle served as the 'target' texts for many commentators, the texts towards which comments were directed. While Simplicius (sixth century CE), in his commentary on the *Categories*, explains that he hopes to make Aristotle's work accessible to others, this does not preclude his giving a Neoplatonic reading of the text, emphasising agreement in the ideas of Aristotle and his predecessor, Plato. Mathematics texts were treated by some commentators, but this did not necessarily mean that those commentators were mathematicians. For example, Proclus' commentary on the Euclidean *Elements* is arguably more concerned with philosophical questions than with geometry.[4] And while commentators may have sought to explain the ideas of

---

[3] See LSJ, s.v. *hupomnēma* 5. c, citing, for example, Galen's distinction between the *sungrammata* of Hippocrates and Galen's own commentaries (*hupomnēmata*) on them (ed. Kühn, vol. 16, 811); Galen also distinguishes *hupomnēmata* (clinical notes) from the *sungrammata* of Hippocrates (ed. Kühn, vol. 16, 532 and 543). LSJ also cites various scholiasts' use of the term to refer to commentaries on Homeric works. The word *hupomnēma* was also sometimes used by ancient authors to refer to a rather different type of text, a brief note or memorandum.

[4] See Mueller 1992; Netz 1999b 302–303.

other authors, their explanations did not always entail agreement with the ideas presented in those texts. Hipparchus (second century BCE), in his commentary on the *Phaenomena of Aratus and Eudoxus*, points to errors in Aratus' work.[5] The Christian John Philoponus (sixth century CE) famously argued against Aristotle in several of his commentaries. Other commentators also disagreed with the authors of their target texts, though perhaps not so fundamentally.

In certain instances, commentators were keen to preserve texts which might otherwise have been lost; indeed, some of the works we now have were preserved only through their efforts. Simplicius deliberately preserved fragments of Parmenides; he transcribed extracts of Parmenides' poem into his commentaries on Aristotle's works *On the heavens* (dealing with cosmology and astronomy) and *Physics*, explaining that his motive in copying extracts of the poem was to preserve it, because of its scarcity.[6] In his commentary on the *Sphere and Cylinder* Eutocius preserved twelve solutions to the problem of duplicating the cube.[7]

Commentators played a large role in shaping the way the works of their predecessors were read and regarded, and some texts survived only in the language into which they were translated by the commentator. Calcidius (fourth century CE) wrote a commentary on Plato's *Timaeus*; his Latin translation (to 53c only) represented Plato to Latin readers throughout the Middle Ages. Commentary continued to be a very important genre in the medieval period, used in different cultures and contexts, Christian, Jewish and Islamic.

Not all commentaries were composed for the same type of audience. Different commentators had different motivations and aims in pursuing their work. As already noted, some were produced within pedagogical settings, aimed at students, while others were produced for more advanced readers. A number of important commentaries were produced by specialist astronomers and professional physicians; the audiences for these also varied. Hipparchus' *Commentary on the Phaenomena of Eudoxus and Aratus* targeted one text which was a poem, the other prose, but both were concerned with astronomical phenomena.[8] The physician Galen of

---

[5] Hipparchus 1.2.1-22 (ed. Manitius 1894: 8–23). See also Tueller and Macfarlane 2009: 234–237.
[6] See also discussion in Chapter 1.
[7] It is due to Eutocius' commentary-cum-anthology that we have the text of Eratosthenes' *Letter to King Ptolemy*.
[8] Some scholars, including Toomer 2012, have described Aratus' work as a poetic version of Eudoxus' prose work. Cf. Gee 2013, who argues that Aratus was not simply offering a poetic setting of Eudoxus' prose.

Pergamum (129–216 CE) produced as many as seventeen commentaries on Hippocratic works; while he was much concerned with medical matters, Galen also had deeply philosophical interests, and discussed textual matters of scholarship as well.[9] Specialist mathematical texts, including those by Apollonius of Perga and Claudius Ptolemy, were the also subject of commentaries, composed by commentators with a high level of mathematical expertise, including Theon of Alexandria (335–405 CE) and his daughter Hypatia of Alexandria (350–415 CE). Of the numerous commentaries produced on philosophical texts, many of which may be regarded as philosophical works in their own right, some were written for publication (such as those by Alexander of Aphrodisias and Simplicius on works by Aristotle), while others were transcriptions of lectures. In this latter category might be included some commentaries by Philoponus and Olympiodorus (Dickey 2007: 49; Verrycken 1990: 241–242 on Philoponus). Ancient commentators may not have conceived of their works as constituting examples of a uniform genre we would call 'commentary';[10] nevertheless, there are elements of these works that indicate their status as a type of text. In particular, as Ineke Sluiter has pointed out, commentaries are explicit about their 'second-order status' and their 'direct relation with and dependence on an earlier text or author'; commentaries are works of scholarship 'whose explicit purpose is to elucidate a text' (Sluiter 2000: 183). However, as we will see, the authors of commentaries at times strove not only to explicate a text but also to present their own views on subjects, sometimes touched on only tangentially in the target text.

The teaching function of the commentary is often emphasised, and is typically marked out structurally and functionally within the text. Several structural features found within the commentaries on Aristotle's works shed light on the pedagogical design and function of these texts. Working within a tradition in which students were encouraged to produce a written record of the teacher's lectures and add comments, this practice is highlighted in the introductions to various commentaries, and also by the use of the expression 'from the voice of', as when a commentary is presented 'in the voice of Ammonius'; sometimes the name of the student is recorded as well, as when Philoponus notes that he has added 'some

---

[9] In time, Galen's writings themselves became the subject of commentaries by later writers, including Stephanus of Athens (seventh century CE). See von Staden 2002.

[10] The vocabulary used to denote the works we refer to as 'commentary' is not entirely homogeneous and uniform. Furthermore, the structure of ancient commentaries varied; see, e.g., Évrard 1957.

reflections of his own' (in *Posterior Analytics, Generation and Corruption, de Anima*; see Sorabji 1990: 7; also Praechter 1990: 47–49).

While the production of commentaries may have been spurred as part of the developing literary and scholarly culture of the 'book' (see Taub 2000: 29; Roberts and Skeat 1983), such texts were produced and read not only by individual authors and readers but also as a part of a group activity (von Staden 2002: 132–133). In his biography of Plotinus (205–269/270 CE), Porphyry (234–c. 305 CE) offers a glimpse of the culture of the commentary, indicating that numerous commentaries were shared: 'In the meetings of the school he used to have the commentaries read, perhaps of Severus, perhaps of Cronius or Numenius or Gaius or Atticus, and among the Peripatetics of Aspasius, Alexander, Adrastus, and others that were available'. Not only does this list of available commentaries suggest a well-stocked library; Porphyry also makes it clear that reading of the texts was done aloud, and as a group activity (Porphyry *Life of Plotinus* 14.11–15, trans. Armstrong 1966: 41; see also Grafton and Williams 2006: 33–35; Konstan 2007: 4–5; Snyder 2000: 116–117). Philoponus divided his commentaries into portions which could generally be delivered orally within an hour (Évrard 1957: 152–153). The teaching function of Olympiodorus' commentary is reflected in the organisation of the work, which is divided into sections, each representing a *praxis* or lesson.[11] Sluiter has noted that those commentaries which were used as school-texts represent an 'open' genre, which was more liable to be updated than more 'closed' genres (Sluiter 2000: 191–192). And it is important to recognise that not all commentators had a classroom audience in mind: Karl Praechter has noted that Simplicius' commentaries do not seem to have been written as 'lecture' notes; furthermore, Simplicius refers to the activity of reading (or studying) and readers, suggesting that he was writing for a non-classroom-based audience (Simplicius *On Aristotle's On the Heavens* (*de caelo*) ed. Heiberg 1894: 102, line 16; *On Aristotle's Categories* ed. Kalbfleisch 1907: 3, line 14; see also Praechter 1927: column 205; Sorabji 1990: 9).

Most ancient commentaries on philosophical works, including those concerned with the physical world, are organised by *lemmata* (plural of *lemma*), that is, quotations of words or phrases from the 'target' text which is being commented upon.[12] As an example, Philoponus' commentary on Aristotle's *Physics* begins with a proem or prologue in which Philoponus discusses the

---

[11] Praechter 1990: 47–49 discusses the ways in which lessons may have been structured. See also Évrard 1957.
[12] See Sorabji 1990: 8 on the structure of Philoponus' commentaries on Aristotle.

Commentary 91

place of this work within the larger corpus, amongst other topics. Following his introduction, Philoponus directly addresses the Aristotelian text, after quoting the opening passage of the *Physics* (from 184a10):

> Since knowledge and understanding come about in all the disciplines that have principles or causes or elements from knowing these <principles or causes or elements>.

Philoponus then comments:

> It is Aristotle's custom to begin his works from certain common assumptions. Hence in the *Metaphysics* [1.1, 980a21] he started from a common agreement: 'All human beings', he says, 'naturally long to know. The evidence is their love of the senses'; and in the *Posterior Analytics* [1.1, 71a1], 'All teaching and learning come from pre-existing knowledge'; and in the *Ethics* [1.1, 1094a1], 'Every technical or scientific discipline, and similarly every action and choice, seems to aim at some good'.[13]

Emphasising that Aristotle is following his usual practice, Philoponus notes that 'Here too [at the beginning of the *Physics*] in the same way it is from a common assumption that he [Aristotle] makes his beginning'. Philoponus elaborates: 'The assumption is as follows: every science that has principles or causes or elements becomes known when the principles and causes and elements come to be known'. He then goes on to discuss what this means in some detail and at some length (the equivalent of a couple of modern printed pages), before re-quoting the opening lines from the *Physics* ('Since knowledge and understanding come about in all the disciplines...'), offering a further discussion, including a consideration of specific terms used for 'knowledge' (*eidenai*) and 'understanding' (*epistēmē*), how they are used and understood (trans. Osborne 2006: 25; 28).

Commentators use *lemmata* to organise their commentaries, but not all commentators present their discussion in exactly the same way. Philoponus, as did some other commentators, including Olympiodorus, provided a 'double' commentary: first, a preliminary general discussion or *protheoria* is presented following a single lemma, such as we saw in his initial discussion of the opening passage of the *Physics*; a more detailed textual discussion or *exegesis* follows, which may engage with a number of *lemmata* (see Sorabji 1990: 8; Osborne 2006: 7–8).[14] The portion

---

[13] The angular brackets indicate words omitted from the manuscript, and supplied later by an editor. Trans. Osborne 2006: 25.
[14] The way the text is presented in the printed edition in *Commentaria in Aristotelem Graeca* is not always clearly indicative of Philoponus' systematic work as an exegete, as both Sorabji and Osborne indicate.

quoted above is from the *protheoria*; following this, Philoponus dives into his *exegesis*, in which the quotation of an individual *lemma* is, notionally, the focus.

There are many questions that could be raised regarding the aims of commentators, including the preservation of older texts, as well as teaching and explaining the views of others. In the case of Philoponus, putting forward his own views was also an important goal within his commentaries. These various functions reflect, to some extent, differing strategic aims of the commentators. While strategy refers to the aim or goal of the commentator, the choice of tactics – the methods adopted to achieve the strategic goal – was in some cases commonly shared, in other instances more individual and circumstantial.

It is useful to concentrate on a small group of commentaries as a way of comparing and contrasting differences and similarities, to gain understanding of the genre, and its importance as a means of engaging with the views of predecessors while offering opportunities for new insights and ideas. We are in a very nice position with Aristotle's *Meteorology* to compare commentaries on the same text, since we have significant portions of three, those by Alexander of Aphrodisias (*fl. ca.* 198–209 CE), Iohannes (John) Philoponus (*ca.* 490–570s CE, who produced a commentary on *Meteorology* after 529, which has not survived in its entirety) and Olympiodorus (sixth century CE, who produced a commentary on *Meteorology* after 565, which is also not completely extant). There are structural similarities between these commentaries, as well as some differences. Alexander's commentary may have been produced for 'publication,' while those of Philoponus and Olympiodorus may represent their lectures. Both Philoponus and Olympiodorus repeatedly mention Alexander's commentary on Aristotle's *Meteorology* in their own commentaries on that work.

Here, while keeping in mind generic questions relating to the commentaries on Aristotle's *Meteorology*, as well as the various strategic aims pursued by individual commentators on the work, our attention will focus on some of the explanatory tactics employed by Aristotle in the *Meteorology* which play a characteristic role in that text, as we examine the ways in which these tactics are further deployed (or not) by Alexander of Aphrodisias, Philoponus and Olympiodorus. The specific explanatory tactics that will be considered are the use of different sources of knowledge and the use of diagrams. The different ways in which Aristotle's own tactics are adopted and adapted by these commentators underscore the 'openness' of the type of text they were writing, allowing for significant variety and even innovation, while still adhering to conventions of the genre.

Commentaries on Aristotle's Meteorology

The ancient commentators who wrote on Aristotle's *Meteorology* not only explained his work but added their own ideas and criticisms.[15] These commentators on the *Meteorology* targeted other Aristotelian writings as well, and in some cases also wrote works themselves in other formats. Alexander of Aphrodisias wrote commentaries on several texts in the Aristotelian corpus; his commentary on the *Physics* is lost, but those on the *Meteorology* and the *Metaphysics* survive. Philoponus, a student of Ammonius, produced commentaries on Aristotle's *Categories, Analytica, On Generation and Corruption, de Anima*, and *Metaphysics*, as well as the *Meteorology*.[16] Philoponus was a Christian whose theological views closely inform his readings of Plato and Aristotle; this played an important part in his explication of Aristotle's ideas. Another student of Ammonius, Olympiodorus, produced various works, including commentaries on several Platonic dialogues and Aristotle's *Meteorology*. Of the three commentaries on Aristotle's *Meteorology* considered here, Olympiodorus' is the most extensive, but even his commentary is not complete (ending at Book Four, chapter 10 [388a10]). All three of these commentators were publicly supported teachers: Alexander and Olympiodorus were appointed to teach philosophy, while Philoponus was, officially, a grammarian (or philologist).[17] Olympiodorus was 'professor' of philosophy at Alexandria, at a time when the philosophical and theological orientation of the school – as well as its future – was very likely at issue, being under some threat from the Emperor Justinian.[18]

---

[15] The commentary continued to be a very important genre in relation to Aristotle's *Meteorology* well into the medieval period; see Lettinck 1999.

[16] There is disagreement about whether or not Philoponus' commentary on the *Meteorology* was a lecture; Wildberg 2010: 245 believes that it was. See Wildberg 2010: 244–249 (arguing against Évrard 1957) and Verrycken 1990: 240–243 and 258 on dating the *Meteorology* commentary, and whether or not Philoponus had finished teaching. Verrycken believes that Philoponus' commentary was used in conjunction with his own teaching. But the extent of Philoponus' philosophical teaching is not clear. Even though Philoponus was a *grammatikos*, Verrycken has argued that it is possible that Philoponus taught philosophy from 529 onwards. The year 529 is memorable because of the closing, on the order of the Emperor Justinian, of the Neoplatonist school at Athens, and some scholars have suggested that Philoponus saw himself as a possible saviour of the Alexandrian school; see Verrycken 1990: 240 n. 41 for a partial list of those who hold this view. Verrycken 1990: 242–243 argues against this, and sees Philoponus as having been sidelined as a philosopher, devoting the rest of his life to theology, possibly from the early 530s.

[17] The public funding of these posts indicates the value that was placed on education in philosophy, including natural philosophy. On higher education in late antiquity, see Cameron 1967 and Watts 2008.

[18] Cameron 1967: 670–671 has argued persuasively against the commonly held view that the Alexandrian school was entirely Christianised in the second and third decades of the sixth century.

Aristotle's Explanatory Tactics in the *Meteorology*

In the *Meteorology*, Aristotle employed a variety of explanatory tactics, ways to support and bolster his explanations of phenomena. These explanatory tactics include the use of different types of sources and authorities (including those presented *via* doxography),[19] the use of empirical evidence and observation, the use of analogy, and also the use of what we might call 'experiment' – evidence drawn from experience which is contrived for the purpose. Aristotle regards all of these avenues of information as being available to non-specialists, and, importantly, they reflect familiar experience. Another sort of explanatory tactic – the use of visual material to supplement the verbal explanation, particularly in the form of diagrams – is occasionally employed; this may point to specific sorts of audiences. The references to diagrammatic illustrations may indicate that such images were used as visual aids in a teaching setting; furthermore, particularly in the case of lettered diagrams, this material suggests an audience with some level of familiarity with mathematical texts and explanations (Taub 2003: 103–114). While typically employing a range of tactics, our commentators all relied on a variety of sources and authorities.

Explanatory Tactic 1: References to a Variety of Sources and Authorities

In some of his work, notably his writings on living things and in his collection of political constitutions, Aristotle collated information from a variety of resources. Similarly in the *Meteorology*, Aristotle brings together

---

Ammonius, Olympiodorus' predecessor, had apparently appeased Christian authorities; some compromises may have been adopted so that the teaching seemed less problematic for Christian doctrine. Unlike Philoponus, Olympiodorus was not a Christian, although he identified as a monotheist and held that God is the First Cause (Olympiodorus *Commentary on Plato's Gorgias* 4.3 (trans. Jackson, Lycos and Tarrant 1998: 86; ed. Norvin 1936: 28 [449d24-26]); see Westerink 1990: 332 who suggests Olympiodorus was 'only trying to make his Pantheon acceptable for his Christian students'; see also Griffin 2015: 2–7. See also Praechter 1990: 35–36.

[19] A body of writings which were collections of opinions have traditionally been described by classicists as the '*placita* literature' (opinions = *doxai* or *placita* = 'what it pleases someone to think'); one prominent example of the 'genre' is the second-century CE work falsely identified as Plutarch known as *Five Books on the Placita of the Philosophers concerning Physical Doctrines* (Pseudo-Plutarch *Placita philosophorum*). In 1879, the German classicist and historian of ancient philosophy and technology, Hermann Diels (1848–1922), published a major work, *Doxographi Graeci* ('Greek Doxographers'). This work signals his invention of the terms 'doxography' and 'doxographer'. There were no such terms in antiquity, nor does the term 'doxography' refer to a category of writing explicitly referred to by the ancients. It is not entirely clear how far Diels intended the notion of doxography to extend further beyond this literature. See Mansfeld 1999: 17–19 on the use of the term 'doxography'; Runia 1999: 36; Mejer 1978: 81–82; Taub 2017; Zhmud 2001.

Explanatory Tactic 1: References to a Variety of Sources and Authorities 95

data and examples derived from numerous sources. Within his meteorological explanations, he intersperses specific details about conditions in different places. Aristotle does not claim to have collected all the information and data reported in the *Meteorology* himself; he makes it clear he is relying on information and reports from others, transmitted in various ways. For specific rarely occurring and unusual phenomena, he would have had to use information gathered and reported by other observers; in other works, notably the *History of Animals*, he also relied on such reports.[20] The *Meteorology* contains an accumulation of information collected from earlier natural philosophers, the poets and from shared experience. Aristotle's use of *endoxa* – reputable opinions collected of others, as well as data reported by others – indicates that he regards science as a cumulative group enterprise, in which the work of others in the community (predecessors or contemporaries) is shared and contributes to the larger effort to understand (Freeland 1990: 78).[21]

Similarly, and perhaps unsurprisingly, we see this approach – referring to a variety of sources and authorities – adopted by the commentators on the *Meteorology* as well. These references made by the commentators to the ideas of others reflect Aristotle's own practice in citing others. However, as we will see, the sources cited by the commentators do not simply mirror those mentioned by Aristotle, nor do they map neatly from one commentator to another. Furthermore, the passages within the commentaries where specific sources are cited do not neatly reflect Aristotle's own use of these sources. Individual commentators sometimes cite sources not named by Aristotle. And even when citing the same sources used by Aristotle, commentators do not always cite these with reference to the same point as did Aristotle. These divergences exemplify the 'open' character of the genre of commentary. Even a few examples indicate the ways in which the commentators incorporate and build upon the practice of Aristotle, even as there are differences from his – and other commentators' – explanatory approaches.

---

[20] At numerous points in the *History of Animals*, Aristotle provides information that appears to come from reports from others: for example, statements about the treatment of animals in Egypt (608b30-609a2); the claim that a particular kind of owsel is found on Cyllene in Arcadia, but nowhere else (at 617a13-4); the claim that the blue-bird is common in Nisyros (617a23-4). At 633a16-29, he quotes Aeschylus on the appearance of the hoopoe. (I have relied on Thompson's translation for the names of birds.)

[21] Elsewhere, Aristotle (*Sophistical Refutations* 34 [183b18-30]) makes it clear that this is his view of the manner in which understanding proceeds in other areas, including rhetoric and dialectic; he indicates that it is his (and others') usual practice to build on the work of predecessors.

For example, as Aristotle had done earlier, Alexander makes reference to the ideas of various earlier authorities at various points, including the philosophers Anaximander, Anaximenes, the Pythagoreans, Democritus, Empedocles, Heraclitus, Anaxagoras and Plato. The mathematician Hippocrates of Chios (*fl.* end of the fifth century BCE) is named numerous times. The medical authority Hippocrates of Cos is also cited; the work *Airs, Waters, Places* is part of the Hippocratic corpus.

Poets are also mentioned, in particular Homer and Aratus. To some extent, Alexander is following Aristotle's lead here, for Aristotle had mentioned Homer as knowledgeable about the geography of Egypt (*Meteorology* 351b32-352a2) at one point in his discussion of the changing relationship between water and land; Alexander also referred to Homer in his comment on this passage in the *Meteorology* (ed. Hayduck 1899: 61, lines 23–26). Later in this same discussion, Aristotle made reference to the Selloi, who he notes are now called Hellenes, but were once known as Greeks (*Graikoi*) (352b1-3). Aristotle makes no reference to Homer at this point, but Alexander, in his commentary (351a19, Hayduck 1899: 62, lines 21–24), offers a line from the *Iliad* (16.234) describing the Selloi as sleeping on the ground with unwashed feet. By quoting what he regards as a relevant passage from Homer, Alexander further elucidates the Aristotelian text and displays his own erudition. This brief example of a reference to a passage in the Homeric poems not mentioned directly by Aristotle is indicative both of the continued importance of Homer to later scholars, including those writing on natural philosophical topics, and of the willingness – indeed the desire – of commentators to offer their audiences something fresh, even if some of what is newly proffered are quotations from the most ancient poets.

To some extent, Philoponus employed this same tactic, making reference to some of the same philosophical and poetic writers, but not limiting himself to the same sources or examples used by Aristotle and then Alexander. Philoponus also refers to texts not cited by either by Aristotle or Alexander in the context of meteorology, such as Plato's *Timaeus* (referred to by title at 3, 26), presumably presented as a marker of his commitment to some aspect of Platonic physics (cf. 32, 29). Like both Aristotle and Alexander, Philoponus mentions the ideas of earlier philosophers, including the Pythagoreans, Democritus, Empedocles and Anaxagoras. Quotations from several poets are also included, among them Homer (both the *Iliad* and the *Odyssey*), Hesiod (the *Theogony*) and Aratus (from the section of the *Phaenomena* dealing with signs). Philoponus broadens the scope of references to literary works by others, including Euripides

Explanatory Tactic 1: References to a Variety of Sources and Authorities 97

(fifth century BCE)[22] and Apollonius Rhodius (third century BCE), the author of the *Argonautica*, neither of whom was cited by either Aristotle (whose death predates Apollonius Rhodius) or Alexander.

While Philoponus cites some of the same authorities and sources used by Aristotle and Alexander, there are intriguing differences. For example, in his comment on Aristotle's treatment of other thinkers' views about the Milky Way, Philoponus discusses (at the beginning of 1.8) the poetic and mythic character of some of the Pythagorean material cited by Aristotle. He first offers the *lemma*, quoting the *Meteorology* at 345a13-18 (101, 20):

> Now, some of the so called Pythagoreans say that it [the Milky Way] is the path of one of the stars fallen out at the time of Phaethon's legendary death, others say that the sun was once carried along this circle, in such a way that this place has been completely burnt out or has been affected in some other way by its [the sun's] motion.[23]

Philoponus then (101, 24) explicates the text, indicating that the story was well known:

> The two opinions set forth earlier which are fabulous and made up by poets are in the tradition of Pythagorean myths. The fiction about Phaethon is celebrated in many places in poetry ... (Trans. Kupreeva 2012: 85–86.)

Seemingly criticising Aristotle for not taking these myths seriously enough, Philoponus not only discusses 'natural' explanations of the Milky Way but also offers further details about the myths, including a quotation of a passage from Euripides (*Orestes* 1001–1006), whom he cites by name. With reference to the myth about Phaethon, he suggests that '[i]t is possible that at the time in question many comets, shooting stars and meteors were formed, from which the myth not unlikely had its origin' (102, 35; trans. Kupreeva 2012: 87). Philoponus seeks to explain the myth, as well as the phenomena. He treats the Milky Way at length, returning to the question of mythological explanations. He chastises Damascius (480–*c*. 550), who he notes regards as fact Empedotimus' account of the Milky Way – suggesting that souls follow it as a path on their way to heavenly Hades – rather than treating it as a mythological story.[24] Philoponus

---

[22] For example Philoponus, ed. Hayduck 1901: 102, lines 1–3 (quoting from the *Orestes* 1001–1006) and 19, line 24 (quoting fragment 998, associated with Euripides, ed. Nauck 1902: 278–279).

[23] As Kupreeva 2012: 135 n. 274 points out, Aristotle's text in most modern editions has 'by their motion' (*autōn*), which Lee 1952: 58–59 reads as 'of these bodies'; see Thillet 2008: 107 'of these stars' ('ces astres'). See also her comment (Kupreeva 2012: 31–32) on Philoponus' text of Aristotle, for this passage.

[24] Philoponus may have been referring to a (now lost) commentary on the *Meteorology* by Damascius (Kupreeva 2012: 2–3; see also Combès 1986: xxxix–xli). Heraclides of Pontus (387–312 BCE) is the

complains that Damascius, '[j]ust as he was well aware of the objections to Aristotle's theory ... should also have examined each point in the myth, whether it belongs to the nature of things'. For, as Philoponus emphasises, natural things should be explained by natural causes.

Olympiodorus, like the other commentators, refers to the works of a range of earlier authors, citing Homer extensively, as well as Hesiod (*Works and Days*),[25] and Aratus. As did Philoponus, he made reference to the writings of Apollonius Rhodius and Euripides.[26] He cites Theophrastus, as well as earlier thinkers, including Anaxagoras, Democritus, Empedocles, the Pythagoreans and Plato (*Phaedo* and *Timaeus*, as well as other of Plato's works not cited by the other commentators). Olympiodorus refers to Alexander's commentary on the *Meteorology*, but not to Philoponus'.

Both Alexander and Philoponus referred to the work of the mathematician Hippocrates of Chios; in addition to Hippocrates, Alexander and Olympiodorus also mention Euclid, while Philoponus and Olympiodorus mention Ptolemy. Obviously, chronology plays a role here (one can only refer to predecessors and contemporaries), but that is not the only factor determining whether a particular source or authority is named.[27]

Aristotle himself, and his commentators (including Philoponus), discussed and sometimes rejected various etymologies of words as part of their arguments to support their views.[28] Olympiodorus used the investigation of the etymology of particular words as part of his explanatory approach, and does so in a manner which refers to specific sources and types of authorities. For example, in his discussion of whirlwinds (typhoon), Olympiodorus explains that this type of sudden storm:

> Homer calls *Thuella*, but Aristotle Typhon, in consequence of vehemently striking against as it were and breaking solid bodies. Sailors however call it Syphon, because like a syphon it draws upward the water of the sea. But the Alexandrians call it in their native tongue *anemosoure*, because it resembles the circular bed-chambers of women, which the inhabitants call

---

source of Empedotimus' account, and Empedotimus may himself have been a fictional creation of Heraclides (see Kupreeva 2012: 141 n. 362).

[25] For the latter, see Olympiodorus, in ed. Stüve 1900: 112, line 34.

[26] For example, Olympiodorus mentions Apollonius and his *Argonautica* (ed. Stüve 1900: 105, line 10); Euripides is also mentioned (ed. Stüve 1900: 129, 24).

[27] This would be a fruitful area for further study.

[28] See, for example, their comments on the etymology of *aither*: Aristotle *Meteorology* 339b16-27; Aristotle *On the Heavens* 270b19-25; Philoponus ed. Hayduck 1901: 16, trans. Kupreeva 2011: 44–45. See Plato *Cratylus* 410b on the etymology of *aither*. See also Wildberg 1988: 9–12.

anemosouris. And physicians from the similitude of the passion produced in the spirals of the intestines, call it *borborugmos*.[29]

Here, Olympiodorus presents an unusually wide range of sources in one passage. He mentions the names used by, arguably, the most influential epic poet (Homer) and the philosopher who wrote most systematically on meteorology (Aristotle). But he does not confine himself to the terminology used by these two central and illustrious figures. Rather, his list of terms employed for whirlwinds includes words used by ordinary people, the inhabitants of his own hometown, Alexandria, as well as words used by professional specialists (sailors and physicians) whose livelihoods, to varying extents, were based on knowledge and understanding of the effects of meteorological phenomena. In other words, Olympiodorus has a wide and diverse range of authorities he is willing to cite, including the 'Alexandrian-in-the-agora', as well as expert professionals. Aristotle had indicated at numerous points his own debt to – often anonymous – others for providing meteorological information, for example, about the occurrence of a particular meteorological event – such as the 'recent' earthquakes he mentions in Heracleia and Hiera (366b32-367a9) – as well as possible explanations. While Olympiodorus' commentary was learned – as evidenced by deliberate and, perhaps, expected – citations of Homer, some of the 'learning' he incorporated was acquired from a broad range of people, and was not based only on older philosophical texts, but instead on contemporary native informants in Alexandria.

The comparison of the range and types of sources and authorities cited by Aristotle and his commentators shows that this was not a static explanatory tactic. The 'openness' of the genre of commentary supports this variety. While there may have been an expectation that Homer would be called upon as a 'witness' giving testimony or evidence to support an explanation, as we have seen above, even Homer was used in different ways; it was not always the same passages being quoted or referred to in the same contexts relevant to the target text. Commentators had the opportunity to display their erudition, for example, by citing a range of authors and types of sources, including dramatists, mathematicians and physicians, and also to update their information, in some cases by

---

[29] Olympiodorus 3.1 (*Meteorology* 370b3; Stüve 1900: 200), trans. Taylor 1807: 540, slightly altered. Taylor explains that the word *borborugmos* is meant to refer to the noise of the intestines; this is what we might call 'tummy-rumbling'. The *OED* provides examples of 'borborygm' in modern English usage, including in works by Erasmus Darwin and H.G. Wells (www.oed.com/view/Entry/21605?redirectedFrom=borborygm#eid; accessed 28 February 2016).

citing sources which post-date Aristotle. Commentators also cite their predecessor-commentators. These citations of different sources may give an indication of the interests and expertise of each of the individual commentators, distinguishing their treatments from others; these differences may also be indicative of different target audiences and their expectations. The variations displayed in citing different sources and authorities underscore our understanding of commentary as an open genre, allowing – perhaps even encouraging – diverse responses to the target text, and offering commentators scope for their own interests and creativity.

Explanatory Tactic 2: Using Diagrams

Another intriguing explanatory tactic found in the *Meteorology* is the reference to, and implied use of, diagrams and illustrations described there, which are now no longer extant. For example, Aristotle refers to lettered diagrams in his explanations of several phenomena, notably haloes and rainbows. He also refers to a lettered diagram in his discussion of the positions of the wind.

While Aristotle referred to diagrams to help explain these phenomena, his commentators vary in when they choose to refer to diagrams. For example, very detailed descriptions of diagrams are part of Alexander's treatment of the winds (363a8), and the rainbow and haloes (373a4, 375a1, 375b16, 375b30, 376a10, 376b28), as they were for Aristotle. In contrast, in Aristotle's discussion of the habitable sectors of the earth (at 362a32), there is no explicit reference to a diagram, even though he speaks about the earth's surface, and describes drum-shaped sectors and 'cones'. Yet Alexander, in his discussion of the habitable sectors of the earth, does refer to a lettered diagram (362a32), as does Olympiodorus, who describes a diagram relating to the habitable sectors of the earth somewhat later in his own commentary (362b1 and 362b5); his description is very brief. Philoponus also uses geometrical diagrams, and in particular his inclusion of diagrams may indicate some of the ways in which he places his individual stamp on subjects he is explicating and explaining.

However, before considering a specific example, it is useful to think more generally about lettered diagrams in Aristotle's works. There are detailed references to lettered diagrams at several points in the *Meteorology* and it is very likely that Aristotle used diagrams elsewhere, for example, in his anatomical, physiological and zoological studies. While, as in the *Meteorology*, the drawings themselves do not survive, at certain points in his zoological works, for example, in the *History of Animals*, Aristotle refers to diagrams (for

Explanatory Tactic 2: Using Diagrams

example 497a32, 525a8-9.; see also Lloyd 1978: 230–231 n. 7). It has sometimes been suggested that the diagrams referred to in Aristotle's works may have served as visual aids for his lectures (see Lee 1952: 67 n. *b*; Jackson 1920; Netz 1999a: 37 n. 65).[30] A lettered diagram is referred to in the *Historia Animalium* 3.1, in a context which has nothing to do with mathematics.

While there may be no known prior examples of lettered diagrams, scholars have argued that earlier authors, including Anaxagoras and some Pythagoreans, used diagrams in their works too.[31] A particular passage in Iamblichus' *On General Mathematical Science* XXV has received a good deal of attention, emendation and re-translation, and indicates that the use of diagrams was an important feature of the Pythagorean approach to studying optics:

> The Pythagoreans devoted themselves to mathematics and both admired the accuracy of its arguments, because it alone, of those things which humans pursue, admitted of proofs, and also saw that the study of harmonics was in a state of agreement, because it was based on numbers, no less than the study of optics, because it was based on diagrams [*grammatōn*].[32]

A number of scholars have argued that this passage in Iamblichus preserves a fragment from Aristotle's discussion of Pythagorean optics (Burnyeat 2005; Huffman 2005: 563–566; Burkert 1972: 50 n. 112).[33] Rather than regarding diagrams only as visual aids for lecture, Aristotle may have accepted what seems to have been the Pythagorean position that the study of optics required the use of diagrams; diagrams may thus have been an essential feature of the presentation of the argument.[34]

---

[30] There are some scholars who suggest that Aristotle may have referred to diagrams in the *Prior Analytics* (Kneale and Kneale 1962: 71–72; Greaves 2002: 116–117; Netz 1999a; see also Fowler 1999: 389). This also relates to the wider question of the relationship between Aristotle's syllogistic and mathematics, a question which cannot be examined in detail here but is nevertheless relevant to the understanding of the use of diagrams more generally in presenting arguments. There is a considerable literature on this topic; see particularly Frede 1974; Mueller 1974; Smith 1977 and 1989 (e.g. on 25a5–13; 26b21; 27a14–15; 42b8); Fowler 1999: 388–390.

[31] Netz 1999a: 61 has suggested that there are no obvious precedents for Aristotle's practice of using the lettered diagram. See Sider 2005: 15–19 on diagrams in ancient texts, particularly those of Anaxagoras; Burnyeat 2005 and Huffman 2005: 552–569 on Pythagorean work on optics, citing Iamblichus *On General Mathematical Science* XXV.

[32] Iamblichus *On General Mathematical Science* XXV, trans. Huffman 2005: 552–553. See also Burkert 1972.

[33] Burnyeat 2005 has argued that Plato was opposed to the Pythagorean approach to optics, and to the use of mathematics to explain the physical world.

[34] That the word *diagramma* may sometimes have been used in mathematical writings to refer to a 'proof' or 'theorem' rather than a diagram is not disputed; however, it is also clear that in some philosophical texts the word referred to visual diagrams (for example, Plato *Republic* 529d-e). See Netz 1999a: 36 on use of *diagramma* to refer to 'proof'; Huffman 2005: 566.

In the passage quoted above, Iamblichus indicated that, for some, diagrams played a key role in setting out an argument. But figures, including diagrams, may not necessarily be tied to demonstration; in some instances, a figure (a *diagramma* or *hupographē*) may be used to explicate – rather than to strictly demonstrate – an argument. Figures may help to make a point evident for a particular case, or may indicate links along which an argument may proceed, by analogy, as it were. In some cases, arguably, such figures take the reader beyond the text.

As noted earlier, there are a number of diagrams and illustrations described by Aristotle in the *Meteorology*. For example, in his discussion of the Milky Way in Book 1, Aristotle (346a32) refers to a *hupographē* which illustrates the stars: 'The circle and constellations in it may be seen in the *hupographē*'. While the character of this *hupographē* is not certain, the reference must be to an illustration of some sort, presumably a drawing or a diagram. This *hupographē* does not appear to be a lettered mathematical diagram.[35] None of our three commentators on the *Meteorology* – Alexander, Philoponus and Olympiodorus – refers to a *hupographe* or diagram in their discussion of the Milky Way.

The term *hupographē* is also used to describe the depiction of wind direction, via a sort of 'wind-rose', the construction of which Aristotle details in Book 2.[36] This *hupographē* (at 363a26) is offered as an aid to following his exposition of the positions of the winds and the description of those that can blow simultaneously, as well as their names and number (363a21-364a4.). Aristotle's opening words indicate the function of the diagram in this passage, in relation to the text: it is intended to illustrate and augment the verbal description, and the illustration is described in sufficient detail to allow it to be reproduced. The description begins as follows: 'For the sake of clarity we have drawn the circle of the horizon; that is why our figure is round. And it must be supposed to represent the section of the earth's surface in which we live; for the other section could be divided in a similar way' (*Meteorology* 363a27-29; trans. Lee 1952: 189). Furthermore, this *hupographē* is a lettered diagram, which makes reference to the physical world.[37] Yet, the description of the *hupographē* relating to wind position has something of a mathematical flavour, incorporating

---

[35] The use of lettered diagrams is sometimes regarded as characteristic of Greek geometrical texts; see Netz 1999a.

[36] Aristotle was interested in the problems of graphical representation: this is made clear by his criticism of the way in which contemporaries drew and mapped the earth, a subject he touches on in his treatment of the winds (362b12-15). See also Harley and Woodward 1987: 145.

[37] The diagrams in Book 3 that describe rainbows and haloes also make reference to the physical world. In Book 1, the term *hupographē* refers to an illustration without letters, in Book 2 to a lettered diagram.

Explanatory Tactic 2: Using Diagrams

technical formulaic geometrical language, as in this extract (*Meteorology* 363a34-b7; Lee 1952: 189):

> Let the point A be the equinoctial sunset, and the point B its opposite, the equinoctial sunrise. Let another diameter cut this at right angles, and let the point H on this be the north and its diametrical opposite Θ be the south. Let the point Z be the summer sunrise, the point E the summer sunset, the point Δ the winter sunrise, the point Γ the winter sunset. And from Z let the diameter be drawn to Γ, from Δ to E.

Here Aristotle employs technical terms (for example, *diametros* = 'diameter') and he refers to letters in his construction. But in spite of the seemingly mathematical features of the text and the description of the lettered diagram (depicted in Figure 4.1), the explanation and discussion are not purely mathematical – that is, the text is not about mathematical objects. Rather, it is meant to define the positions of the winds. The word translated here as 'point' is *topos*, rather than *stigmē* or *sēmeion*, other terms which Aristotle uses to designate a 'point'; the choice of the term *topos* indicates that the language here is not purely geometrical, but also geographical.[38]

A brief look at the commentators shows that both Alexander and Olympiodorus include descriptions of lettered diagrams as part of their discussion of the winds; Philoponus' commentary does not extend to this point.[39]

Alexander, Philoponus and Olympiodorus vary in the extent to which they apply what looks like a mathematical approach – incorporating formulaic language and lettered diagrams employed – to topics in the *Meteorology* for which Aristotle himself did not describe a lettered diagram. One example already mentioned is Alexander's discussion of the habitable sectors of the earth (362a32).[40] Another is found in Philoponus' discussion of the size of the earth in relation to the stars and the heavenly bodies, as part of his explication of Aristotle's discussion of the celestial region (339b30-36). There, Aristotle strongly advocates the value of mathematics, explaining that

> those who maintain that not only the bodies in motion but also the element surrounding them are composed of pure fire, and that the space

---

[38] I am grateful to Bernard Vitrac for emphasizing this, in personal communication.
[39] Alexander describes a diagram, in his discussion relating to 363a8 (ed. Hayduck 1899: 107–110); Aristotle does not refer to a diagram here (at 363a8), but does do so later when discussing the winds (at 363a35). Olympiodorus (ed. Stüve 1900: 185, 15–187, 1) also describes a diagram.
[40] Here (ed. Stüve 1900: 189) Olympiodorus has a comment, but no diagram. He does offer a diagram (ed. Stüve 1900: 186) to correspond to 362a31, describing how the south wind blows from the summer tropic (on which Alexander comments, but describes no accompanying diagram).

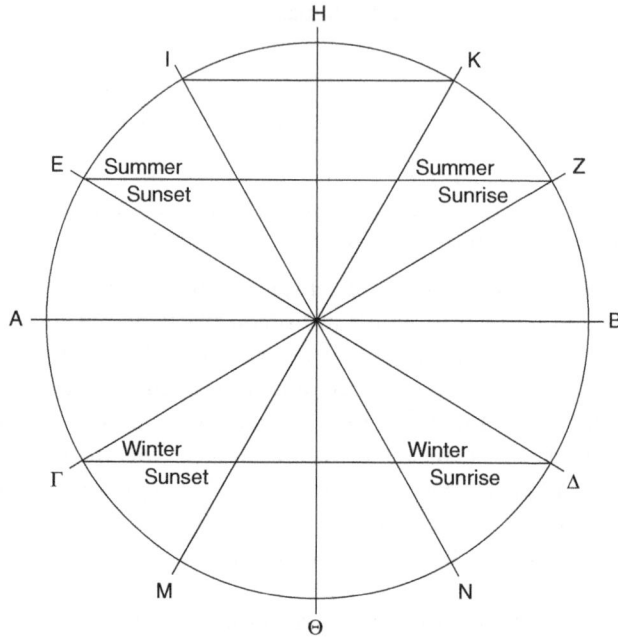

Figure 4.1 Lettered diagram based on the description given by Aristotle in his *Meteorology* 363a34-b7. (The version here is an emendation of that provided by Lee 1952: 187.)

between the earth and the stars is filled by air, would perhaps have ceased to hold this childish opinion if they had studied what mathematics has now sufficiently demonstrated.[41]

He then goes on to argue that 'it is too simple to believe that each of the moving bodies is really small in size because it so appears to us when we look at it from the earth'. Here, Aristotle did not himself use mathematics in his elaboration of his views, nor did Alexander introduce mathematics into his discussion of this passage.

In contrast, in his commentary on this same passage, Philoponus employed geometrical language and mathematics. Philoponus provides a somewhat detailed geometrical discussion as part of his explication of this passage (at 19, 8–30), and uses a lettered *diagramma* as a crucial part of his discussion (at 19, 8–20). The language of the description of the *diagramma* invites the reader to participate in its construction:

---

[41] Trans. Lee 1952: 15.

Let there be a circle ABΓ, and let its centre be Δ and its diameter AE, which necessarily divides the circle into two equal half-circles. Let another circle be described around the same centre, smaller, that is, than the outer one. Let the outer circle be the fixed sphere and the inner circle the earth. Now, assuming that the earth is the surface [of the inner circle], draw from it a straight line ZH to the eastern and to the western horizon. This will divide the fixed sphere into two unequal sections, the one above the earth smaller than a half-circle, let us say, of five signs, and the one underneath the earth of seven, for example, or however it happens to turn out, the one above the earth being ZBH, the one underneath ZΓH.

Now the line through the centre, since the indivisible is in no proportion to the given magnitude, divided the outer circle into two equal parts, while the line through the surface of the inner circle divided it into unequal parts. If, then, the magnitude of the earth were in any proportion to the fixed sphere, the part of the sky above the earth and visible to us would inevitably be smaller than a half-circle. But since at all times we see exactly six signs above the earth, it is clear that, as the centre is in no proportion to the fixed sphere so neither is the earth in any proportion to it, although the earth is quite large and because of its magnitude is called 'land' (*ēpeiros*) as equivalent to 'boundless' (*apeiros*), as in the line, 'We sailed toward a boundless land' [fragment 998, ascribed to Euripides, ed. Nauck 1902: 278–279]. Hence it does not conceal any part of the fixed sphere, any more than the indivisible point does; for if its magnitude were in any proportion to the fixed sphere, we would see the half-circle above the earth to be smaller, as the remaining part that completes the half-circle above the earth would be hidden from sight by the magnitude of the earth, as in the circles of our diagram on the eastern horizon the section A to Z and on the western horizon the section E to H. If, then, the earth is nothing in proportion to the outermost sky, it makes no sense to ask whether it is smaller than some stars, one should rather ask if it is larger than any stars at all.[42]

This is a rather complicated discussion, in which the reference to the constructed *diagramma* is knitted into the argument; whether or not the *diagramma* is, strictly speaking, necessary to the argument is not entirely clear.

And while the mathematical approach is foregrounded (as is shown not only through the lettered diagram and technical terminology but also through the invitation to the reader to construct the diagram), the quotation of a line from Euripides given as an illustration of the meaning

---

[42] Trans. Kupreeva 2011: 48, slightly altered (including the adoption of the letters from the Hayduck edition).

of *apeiros* (unbounded, as in 'We sailed toward a boundless land') is vividly evocative of physical space rather than mathematical abstractions. It also is another opportunity for Philoponus to display his literary erudition. Following the discussion of the size of the earth, and the diagram, Philoponus then turns to the size of the sun, citing the work of anonymous mathematicians, once again showing his familiarity with a range of types of authorities and sources, as well as the breadth of his knowledge.

Elsewhere in his commentary on the *Meteorology*, Philoponus uses a lettered diagram and technical mathematical language to explain particular meteorological phenomena – shooting stars. In his treatment of Aristotle's explanation of shooting stars, 'torches' and 'goats' (342a21 at 66, 4–17), in which Aristotle states that 'the movement of shooting stars is commonly transverse [*loxē*; slanting or cross-wise]', Philoponus offers a rather elaborate discussion of the movement of shooting stars through the sky, complete with a description of a lettered geometrical diagram. However, the portion of the discussion which employs the lettered diagram does not seem to be about geometry (*qua* mathematics), but rather about motion (66, 3; trans. Kupreeva 2012: 48):

> 'For', he says, 'all such things move along the cross-section'. For the cross-section of rectangles is crosswise with respect to the sides by which the rectangle is formed. So, he likens the sideways ejection of the shooting stars to the motion along the cross-section of objects which previously moved along the sides of the rectangle but have been pushed out by each other at the corner by a mutual collision so that they get carried off at a diagonal.

Philoponus explains that Aristotle was concerned with explaining shooting stars and related phenomena as things which are moving. He describes a geometrical shape (rectangle) and uses geometrical terminology (diagonal) to set up what might be described as an analogy to presumably familiar experience, or what may even be a thought experiment, about ants tracking along a path (see Figure 4.2):

> Assume a rectangle ABCD, with its cross-section, i.e. diagonal, AD; let two ants of equal strength be moving, one from C to A and another, again, from B to A. When they are at A and neither gets the better of the other as they push each other, they are shoved off the sides of the rectangle, and being deflected, they get carried off along the cross-section AD. (Trans. Kupreeva 2012: 48)

Philoponus' use of an analogy to common experience is an explanatory tactic often employed by Aristotle; indeed, Aristotle drew an analogy in his discussion of these phenomena to the motion of thrown objects (Taub

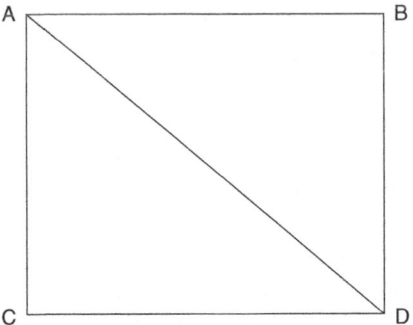

Figure 4.2 Lettered diagram depicting Philoponus' description of ants tracking a path, following Kupreeva 2012: figure 1, p. 48.

2003: 90 (referring to *Meteorology* 342a28-33), cf. p. 101; see also Lloyd 1966: esp. 403–420). But when Philoponus invites us to 'let two ants of equal strength be moving', describing their motion on a plane, he also invokes the language of geometrical proof, when he refers to a lettered diagram, and in setting out the task. Aristotle used references to lettered diagrams as part of his treatment of the winds and the rainbow, but he did not refer to a diagram in his discussion of shooting stars, nor did the other commentators on the *Meteorology*, Alexander and Olympiodorus.

Here, in his discussion of shooting stars, Philoponus is innovating. He employs a technique found in Aristotle and utilised by other commentators, namely the use of geometrical diagrams to explain physical phenomena. Using a familiar technique, Philoponus finds a new application to explain a different phenomenon. This would fit with what Samuel Sambursky (1962: 156) has described as Philoponus' 'often admirable faculty of giving a concrete illustration of an abstract concept or replacing a worn-out analogy by a new and surprising one'. Yet, while Philoponus' explanation is cast in geometrical terms, it is not really a mathematical argument. The rather unexpected use of imaginary ants is a striking example of Philoponus using Aristotle's explanatory tactics (here, both analogy and diagrams) to support a novel and vivid explanation. The analogy between the motion of shooting stars and the motion of ants emphasises that these phenomena occur in the same sort of physical space (see also Wildberg 1988).

As we have seen, Aristotle had advocated the use of mathematics in understanding physical phenomena. In his explanation of the passage described above, Philoponus appears to have taken Aristotle's suggestions seriously, going further even than Aristotle himself. That Philoponus offers a mathematically-tinged explanation – using geometrical language and a lettered diagram – in commenting on a passage in which Aristotle had not felt a similar need should not surprise us, for there is other evidence of Philoponus' strong interest in mathematics, including astronomy; a work on the astrolabe has been ascribed to him (see Neugebauer 1975: vol. 2, 878 n. 7 for details of edition and translations), and he produced a commentary on Nicomachus' *Introduction to Arithmetic*.[43]

However, Philoponus did not always agree with Aristotle, and was concerned to distinguish his own views, even as he sought to explicate Aristotle's *Meteorology*. Philoponus held a number of views absolutely opposed to Aristotle's, and he frequently makes his own point of view clear (see Verrycken 1990: 242, and n. 47). A Christian Neoplatonist, Philoponus repeatedly sought to assert the doctrine of Creation (Verrycken 1990: 264–274). In his commentary on the *Meteorology*, Philoponus attempts to eliminate the fifth element, the *aither*, which is, according to Aristotle, eternal and divine (cf. Westerink 1990: 327). He rejected Aristotle's claim that the world is eternal, because, if the world is eternal, creation is denied.

But while Philoponus' views were at times in sharp contrast to those of Aristotle, he did – like the other commentators on the *Meteorology* – incorporate into his commentary many of the explanatory tactics used by Aristotle in the *Meteorology*, even if they were not always used in precisely the same ways as in the text being explicated. In his commentary on the *Meteorology*, Philoponus employs explanatory tactics that were used by Aristotle himself, including, along with analogy, the citation of different types of authorities (including poets and mathematicians), and the use of lettered diagrams; sometimes these tactics are employed in a single passage, as we have seen in Philoponus' discussion of the size of the earth, where he brings them all to bear.

Philoponus disagreed with Aristotle on some fundamental points regarding the nature of the world, but, as we have seen, he was not the

---

[43] Furthermore, Philoponus used lettered diagrams in other non-mathematical contexts. In his commentary on the *Prior Analytics* (1.27 [Arist. 43b1], ed. Wallies 1905: 274) we encounter a diagram schematising which kinds of conclusions follow from specific kinds of premises. This diagram, later to be called 'asses' bridge' (*pons asinorum*), enabled those studying logic to construct valid syllogisms more easily by helping them to identify the middle term; letters are used to refer to different terms (see Wildberg 2008: section 2). Hamblin 1976: 131 notes that in his commentary on the *Prior Analytics*, 'these rules were elaborated by Alexander of Aphrodisias, who speaks of a "diagram" [ed. Wallies 1883: 406, line 12] and may reasonably be credited with having invented it. The diagram itself is given by John Philoponus'.

only commentator on the *Meteorology* to innovate. All of these commentators refer – as did Aristotle himself – to a range and different types of sources and authorities; each of the commentators introduces some sources which are unique to their own discussion, introducing quotations from 'new' authors.

In what has necessarily been a brief look at tactics of explanation used by Aristotle in his *Meteorology* and adopted by his commentators, we see that these commentators adapted Aristotelian tactics in a number of ways – for example, by introducing different authorities and sources, and by using diagrams to describe phenomena not treated in that way by Aristotle. These adaptations of Aristotle's approaches and methods reinforce our understanding of the genre of commentary as being 'open', allowing and facilitating further explication, updating and innovation.

The genre of commentary offered its users different kinds of opportunities to engage with their target texts and their authors, as well as with their own prospective readers. The commentary has a certain structure, the *lemmata* being perhaps the clearest (but not the only or always present) feature of the genre. The commentary's function – to engage with and elucidate a particular target text – does not preclude the commentators from expressing their own ideas. In fact, the genre invites the commentators to insert their own point of view and voice.[44] Remember, many of the later commentaries were described as being written 'in the voice of' a named individual.

The individual voice of the commentator resounded through the genre, even as, in late antiquity particularly, complex relationships developed between Greco-Roman culture and the new religions of Christianity and, later, Islam. The survival or loss of ancient writings, including those on scientific topics, in many cases depended on the attitudes of individual Christians and Moslems. The preservation of pagan texts relied, in large part, on their perceived value within the ascending cultures of Christianity and Islam, and to some extent the culture of Judaism, with its tradition of rabbinic commentary. Like many other natural philosophical texts, Aristotle's *Meteorology* was the target of numerous medieval commentators, as it had been for ancient commentators.[45] As an example, Alexander

---

[44] My use of the gender-neutral pronoun here is in recognition of Hypatia's work and reputation as a commentator.

[45] The importance of the text is highlighted by the fact that it was the first of Aristotle's works to be translated into Hebrew in the medieval period; this was motivated by the view that the text would shed light on the first book of the Hebrew Bible, *Genesis*. See Fontaine 1995.

of Aphrodisias' commentary on the *Meteorology* was translated into Latin by William of Moerbeke (*c.* 1215–1286).[46] Like William, later commentators on ancient scientific, medical and mathematical works, in several linguistic, religious and cultural traditions, incorporated and made reference to the work of the ancient commentators, once again demonstrating the 'openness' of the genre, allowing the voices of distant and foreign predecessors to continue to be heard and to be reckoned with afresh.

---

[46] See Goyens, De Leemans and Smets 2008 on translations.

5

Biography

> At the start of every philosophical investigation, it is after all the custom, at least for all who are sound-minded, to invoke God. But at the outset of that philosophy rightly believed to be named after the divine Pythagoras, it is surely all the more fitting to do this; for since this philosophy was at first handed down by the gods, it cannot be comprehended without the gods' aid. Moreover, its nobility and greatness exceed human ability to understand it immediately: only when the goodwill of the gods leads the way, can someone with gradual approach slowly appropriate something from it.
>
> Iamblichus *On the Pythagorean Way of Life* 1, trans.
> Dillon and Herschbell 1991: 31.

There was no genre of 'biographies of scientists' in the Greco-Roman world, yet information about natural philosophers and mathematicians was communicated in many ancient works. In some cases, the biographical details provided seem to be presented only incidentally. So, for example, we get snippets about incidents in the life of Thales, often celebrated as the first natural philosopher, from Plato (*Theaetetus* 174a). He relates a story, credited to a female slave, that Thales fell into a well while star-gazing. This claim to intellectual unworldliness is somewhat balanced by the account given by Aristotle (*Politics* 1259a5-21) relating how Thales cornered the market in olive presses in Miletus, thereby creating a monopoly.

As was noted in the Preface to this volume, the 'biography' (*bios* [plural = *bioi*], a writing giving an account of someone's life) was not always a clearly demarcated genre for ancient Greeks and Romans. In particular, there are not distinct boundaries differentiating what we might regard as biographies from genres such as the eulogies or *encomia* used to praise heroes and important citizens.[1] Christopher Pelling (2012: 241) has

---

[1] Harvey 1937/1974: 158 gives the original meaning of encomium (*enkōmion*) as 'a Greek choral hymn ... in celebration, not of a god, but of some man'. He explains that 'the word means a song "at the kōmos" (here the revel at the end of a banquet), and thus suggests a eulogy of the host'.

suggested that ancient biography should be understood as a set of 'overlapping traditions, embracing works of varying form, style, length, and truthfulness'.[2] Indeed, accounts of the lives of particular individuals can be found in a range of ancient writings; some of these also include discussion of the opinions (or *doxai*) of the individual being described.[3] Some *bioi* are also intended to impart ethical or religious teachings. While biography itself was not a rigidly defined genre for the Greeks and Romans, the *bioi* are often linked by the desire to celebrate individuals.

Aristotle is often credited with generating interest in the writing of *bioi* that reflected intellectual and ethical concerns (Pelling 2012; Momigliano 1993: 119). A member of Aristotle's Lyceum, Aristoxenus of Tarentum (born *ca.* 370 BCE), is recognised as the author of *bioi* of at least four philosophers, Pythagoras (born mid-sixth century BCE), Socrates (469–399 BCE), Plato (*ca.* 429–347 BCE) and Archytas (*fl. ca.* 400–350 BCE; also of Tarentum); these may have formed part of a series of *bioi* of philosophers (Barker 2012; also Pelling 2012). A number of authors, including Sotion of Alexandria (active between 200 and 170 BCE), are also deemed to have produced such a series. Such collections of *bioi* became a standard form for presenting intellectual history; 'the "succession" of teachers and pupils was a helpful idiom for explaining influences' (Pelling 2012).[4]

Many of the accounts of the lives of ancient 'scientists' – natural philosophers and mathematicians – reflect these two aims: to celebrate the individual and also to present intellectual history and lineage. These two strands of ancient biographical writing are not entirely separate. More generally, many *bioi* – not only those concerned with natural philosophers and mathematicians – emphasise the heroic character of the individual whose life is being described. Plutarch (born before 50 CE, died after 120), in his *Parallel Lives* of Greek and Roman 'great' men, was particularly interested in ethical and moral matters (see Lamberton 2001: 69–74). Certain later *bioi* – for example, the *Life of Apollonius* by Philostratus (died between 244 and 249 CE) – appear almost hagiographical; Apollonius of Tyana (*ca.* 40 – *ca.* 120 CE) was a Pythagorean sage (see Dzielska 1986;

---

[2] Söderqvist 2007 points to the difficulties in speaking about a homogeneous genre of biographies of scientists in the modern period as well.

[3] Those writings known as 'doxographical' concentrate on presenting opinions, without necessarily providing biographical details of the life of the individual (Mansfeld, 1999: 17–19; Runia 1999); see also note 19 to Chapter 4.

[4] Momigliano 1993: 65–100 considers the role of Aristotle's school in shaping Hellenistic biography; in particular, see chapter 4, 'From Aristotle to the Romans'. While autobiography is not treated here, Galen's *On My Own Opinions* can also be understood as offering an intellectual history.

Flinterman 1995; 2009). For some, marvels credited to Apollonius call to mind miracles performed by Jesus; as a number of scholars have noted, 'the Christian Gospels have points of contact with the Greek tradition, with their charismatic hero and their anecdotal narrative texture' (Pelling 2012; see also Talbert 1977: esp. 25–38).

A 'hero' was a member of a class of beings worshipped by the ancient Greeks. Heroes were generally conceived of as the powerful dead, and understood as forming a class intermediate between the gods and human beings. As Emily Kearns has noted, from the fourth century BCE onwards, there was in practice great variation in the types of honours offered to heroes. She explains that 'at one end of the spectrum it could have a strong resemblance to the offerings given to a dead relative; at the other, it might be barely distinguishable from worship paid to a god'; 'there was a tendency in many parts of the Greek world for mourners to depict the ordinary dead in heroic forms, to call them "hero", and even on occasion to establish regular heroic cult and a priesthood' (Kearns 2012; Kearns 1989).[5]

In 1965, Moses Hadas and Morton Smith published *Heroes and Gods: Spiritual Biographies in Antiquity*, an examination of what they called ancient 'aretalogy'. In their preface, they noted that 'aretalogy is not recognised as a word in our dictionaries, nor is the type of literature it designates treated as a separate genre' (xiii). To some extent, 'aretalogy' could be understood as the celebration of *arete*, that is, virtue. But Hadas and Smith (1965: 3) use the term to describe something more specific: 'a formal account of the remarkable career of an impressive teacher that was used as a basis for moral instruction'.[6] Hadas and Smith were concerned with literary images of a particular kind of person, namely a hero or god. They defended their focus on literary images by arguing that:

> For the effect exerted upon the course of history the authorized image of the hero is more important than his historical personality. It is upon the image rather than the person that reverence is bestowed, whether formally in an organised cult or informally in popular tradition, and it is the cult, formal or informal, that ensures the survival of the image. (Hadas and Smith 1965: 4)

---

[5] Diogenes Laertius 5.91 relates how Heraclides of Pontus attempted to organise heroic honours for himself.

[6] Hadas (in Hadas and Smith) 1965: 60 acknowledged that 'we have no complete text surviving from the past specifically labelled aretalogy'. Some scholars, including Burridge 1992/1995: 17–19, have argued against 'aretalogy' as a separate genre, and prefer to use the term 'aretalogical' as an adjective to describe other literary forms; see also Tiede 1972: 1–13 on the 'problem' of aretalogies, and Talbert 1977: 12–13, who rejects the suggestion that the gospels are examples of aretalogical biography.

Hadas and Smith argue that biography, as a constructed literary image, carries historical significance; they centered their study of ancient biography on aretalogical accounts of heroic teachers as moral exemplars, whose activities and teachings may also have religious import as well as ethical influence. Turning to the *bioi* of ancient natural philosophers and mathematicians, there are often important overlaps between the intellectual histories told through a series of *bioi* of philosophers and the aretalogical accounts that concentrate on a particular heroic teacher.

Bioi of Pythagoras

Pythagoras has often been revered, and numerous natural philosophers, mathematicians and scientists have claimed to be his intellectual heirs (Kahn 2001: 153–172). For example, Kepler wrote in a letter to Galileo dated 13 October 1597 that 'you are following the lead of Plato and Pythagoras, our true masters' (1951: 40–41; cf. 1945: 145).[7] Alfred North Whitehead, in his 1925 Lowell Lectures, published as *Science and the Modern World*, credited Pythagoras with 'founding European philosophy and European mathematics' (Whitehead 1925: 54).

In what follows, three accounts of the life of Pythagoras, surviving from later antiquity, are considered. While a detailed analysis and comparison of these texts is not possible here, they nevertheless expose several different strands of ancient biography. The three 'lives' of Pythagoras can all be loosely dated to about the third century CE, and were written respectively by Diogenes Laertius (dating uncertain, but probably first half of the third century CE), Porphyry (234–c. 305 CE) and Iamblichus (c. 245–c. 325 CE); all three were composed significantly long after the death of their subject (fifth century BCE). The life of Pythagoras was celebrated not only in textual accounts; Samos, his birthplace, honoured its favourite son by representing him on coins, marking his heroic status.

Some ancient accounts of the lives of natural philosophers and mathematicians may be regarded as part of a tradition of writing about heroes who were treated as a sort of divine being. In certain cases, the subject of the *bios* is presented as a moral exemplar, whose life and teachings provide a guide for living. Significantly, certain biographies of natural philosophers and mathematicians combine elements that are ethical and/or religious with intellectual history.

---

[7] See also Hallyn 1990: 62 n. 25 (text of the note is on pp. 304–305) on the assessment of the importance of Pythagoreanism to Copernicus, and Jardine and Segonds 1999: 222–226 with regard to Kepler's interest in Pythagoreanism.

Bioi *of Pythagoras* 115

The complicated nature of ancient biographies of 'scientists' – *bioi* concerned with natural philosophers and mathematicians – is especially apparent when we look at the biographies of Pythagoras. In 1962, in his pathbreaking book, *Weisheit und Wissenschaft: Studien zu Pythagoras, Philolaos und Platon*, Walter Burkert argued that Plato and his school had largely created a conception of Pythagorean philosophy (known to us through late ancient sources) to equip themselves with an intellectual heritage (cf. Kahn 2001: 2–3). According to Burkert, much of what we think we know about Pythagoras and those whom Aristotle referred to as 'the so-called Pythagoreans' is, arguably, an example of the desire to construct an intellectual lineage (Burkert 1962; trans. Minar 1972, e.g., 83–96).

Pythagoreanism had two rather distinct forms, or schools, after the fifth century BCE. The 'scientific' or philosophical form (whose advocates were the so-called *mathematikoi*) manifested itself in the fourth century BCE in the thinking of Philolaus and Archytas of Tarentum and the Pythagoreans whom Plato knew and succeeded.[8] The other was a religious, sectarian form, whose adherents were known as *akousmatikoi*, those following certain oral teachings (Graf 2012; cf. Burkert 1972: 192–208). These two strands of Pythagoreanism, the scientific/mathematical and religious/ethical, are both reflected in the extant biographies of Pythagoras, and so make the untangling of these two traditions difficult. Indeed, to attempt to dissociate the two traditions is probably a mistake, for this may give a skewed view of the range of teachings that were offered under the umbrella of Pythagoreanism. These two traditions may also, to some extent, mirror the somewhat overlapping desires of biographers of Pythagoras to produce both intellectual-historical and aretalogical accounts.

Notably, each of the three extant 'lives' of Pythagoras was presented as part of a larger work: Diogenes Laertius' *Lives of Eminent Philosophers*, Porphyry's history of philosophy (*Philosophical History*) and Iamblichus' compendium, *On Pythagoreanism*.[9] Diogenes Laertius' biography of Pythagoras is one of a large number of *bioi* he included in his *Lives of Eminent Philosophers*. Porphyry was a scholar who wrote on a wide variety of topics, including philosophy, grammar, rhetoric and religions. He produced more than sixty works, including commentaries on works by Plato (for example, the *Timaeus*) and Aristotle (such as the *Physics*), as well as *Homeric Enquiries*. Porphyry wrote a history of philosophy from Homer to Plato. His *Life of Pythagoras* is an excerpt from this work (Smith 2012;

---

[8] On Plato's relationship to the Pythagoreans, see Kahn 2001: 39–62.
[9] The relationship between the three works has been the topic of some scholarly debate; see, for example, Burkert 1972: 97–109.

O'Meara 1989: 25–29).¹⁰ Iamblichus probably studied with Porphyry in Rome or Sicily before founding his own school, possibly in Apamea (in modern-day Syria). Many of his writings are now lost, including commentaries on works by Plato and Aristotle. He compiled a compendium of Pythagorean philosophy, incorporating extracts derived from earlier writers; the first four books survive, amongst which *On the Pythagorean Life* is the first (O'Meara 2012).¹¹ These three surviving accounts of Pythagoras' life incorporate material from earlier sources. The first 'life' of Pythagoras was apparently the one written by Aristoxenus (now lost); Charles Kahn (2001: 69) has suggested that many of the marvellous and moralistic features found in the later accounts by Diogenes Laertius, Porphyry and Iamblichus find their source in his work.

Diogenes Laertius' *Bios* of Pythagoras

Any discussion of ancient biographies of 'scientists' must include Diogenes Laertius, because of his *Lives of Eminent Philosophers*. Almost nothing is known about the man himself, including his dates, where he lived, and with whom he studied; he is often disparaged as a 'late' source, but he is our only source of information for many of the philosophers about whom he wrote. His work, probably written in the first half of the third century CE, appears to predate the other extant biographies of Pythagoras.

In his *Lives*, Diogenes Laertius provides biographies and summaries of the opinions of the ancient Greek philosophers, beginning with Thales (sixth century BCE) and ending with Epicurus (341–270 BCE). Diogenes Laertius' work incorporates two approaches to biography distinguished earlier: it provides an intellectual history, but is also motivated by an ethical desire to provide exemplars of the good life (Gardiner 2003; see also Mejer 1978: 2–4). He makes it clear that Epicurus is to be regarded as a model; the virtuous and moral message of this 'life' is explicit. But, as we read Diogenes Laertius' account, we also get a detailed look at Epicurus' natural philosophy; in fact, Diogenes Laertius is responsible for

---

[10] Porphyry was responsible for arranging the work of his teacher Plotinus, into the sets of nine treatises known as the *Enneads*. His treatise on vegetarianism, *On Abstinence*, is well known, but his own writings on metaphysics are almost entirely lost. Of particular relevance to histories of science are the fragments of his commentaries on Plato's *Timaeus*, Aristotle's *Physics*, Ptolemy's *Harmonics*, as well as his introduction to Ptolemy's *Tetrabiblos*, and a treatise on the entry of the soul into the embryo (a work formerly attributed to Galen but probably by Porphyry).

[11] The other surviving books of the compendium are the *Protrepticus* (believed to contain material from Aristotle's lost work by the same name), *On General Mathematical Science* and *On Nicomachus' Arithmetical Introduction*.

the preservation of much of what we know of Epicurus' ideas, through the quotation of three of his letters. In other words, Diogenes Laertius' account of Epicurus is concerned not only with providing an account of his life but also with the details of his work.

Some of the scholars who have worked on ancient *bioi* have examined their structure, arguing that certain features seem to be 'standard'. Richard A. Burridge, as part of his study of the Christian gospels and Greco-Roman biography, identified the following as the usual topics covered: ancestry, birth, boyhood and education, great deeds, virtues, death and consequences (Burridge 1992/1995: 145–147, 178–180, 207–209). An earlier scholar, Armand Delatte, identified general areas according to which Diogenes Laertius organised the biographical material on each of his subjects. Delatte recognised that Diogenes Laertius was a biographer specifically interested in reporting his subjects' intellectual role and contributions; he described the following as the usual topics covered (Delatte 1922: 54–63; cited by Long 1925: xxi–xxii): origin, education (including philosophical training and travels), place in a succession or founding of a school, character and temperament (illustrated by anecdotes and sayings), important life events, anecdotes relating to the subject's death and epigrams, chronological data, works, doctrines, documents (e.g., last will, letters), other men of the same name, and miscellaneous notes (including lists of followers, inventions and political activity). The order of topics identified by Delatte only loosely applies to Diogenes Laertius' *bios* of Pythagoras; nevertheless, there is almost a sense that Diogenes Laertius was following a checklist of themes to be covered. And, indeed, the founding of a school and one's ideas, teachings and writings are particularly relevant for intellectual history.

Diogenes Laertius organises the lives of the philosophers into two 'successions', a method of organisation developed by Theophrastus as well as Sotion of Alexandria (Long 1925/1972: xx; Mejer 1978: 62–74). The first 'succession' is the Ionian, beginning with Thales, continuing through Socrates, and then branching into three 'schools'. The 'Italian' succession begins with Pherecydes and continues through his student, Pythagoras (who moved to Croton, in Magna Graecia, present-day southern Italy).[12]

Diogenes Laertius' *bios* of Pythagoras appears at the beginning of his examination of the philosophy of Italy (Book 8), having ended his treatment of the 'Ionian' succession with his discussion of the philosopher

---

[12] In his Prologue in Book 1 (1.15), Diogenes Laertius names Pherecydes as the first in the Italian school, followed by Pythagoras, and then his son Telauges. We are told elsewhere (8.2) that Pythagoras was a pupil of Pherecydes of Syros.

Chrysippus (c. 280–207 BCE; treated in Book 7). The opening passage reads as follows:[13]

> Having now completed our account of the philosophy of Ionia starting with Thales, as well as of its chief representatives, let us proceed to examine the philosophy of Italy, which was started by Pythagoras, a son of the gem-engraver Mnesarchus, and according to Hermippus, a Samian, or, according to Aristoxenus, a Tyrrhenian from one of those islands which the Athenians held after clearing them of their Tyrrhenian inhabitants (Diogenes Laertius, trans. Hicks 1925: vol. 2, 321).

Here, as elsewhere, Diogenes Laertius names different sources and points to differences in their accounts. (Throughout the *Lives*, Diogenes Laertius proudly names his sources, which number well over 200 [Hope 1930: 59–60; see also Long 1925: xix]; the breaks and shifts that occur at various points in his account are very likely due to his moving from one source to another.)

Continuing, he briefly recounts (8.2–3) Pythagoras' education and travels (which offered him opportunities for study abroad), then reports (8.4–5) what Heraclides of Pontus (fourth century BCE) claimed Pythagoras used to say about himself, regarding his former lives; Pythagoras believed in reincarnation.

Diogenes Laertius then considers Pythagoras' activities as an author (8.6–9), noting that 'there are some who insist, absurdly enough, that Pythagoras left no writings whatever'; he points to evidence, credited to Heraclitus (*fl. ca.* 500 BCE), that Pythagoras did write a number of works (1925: vol. 2, 325). Amongst the several works said to have been written by Pythagoras was one with a title shared by other writings attributed to Presocratic authors – that is, *On Nature*.

Here, Diogenes Laertius emphasises Pythagoras' identification and standing as a physicist (*physikos*). Furthermore, the title of the work, *On Nature*, was that which would have been expected of physicists and, therefore, attests to Pythagoras' reputation as someone who philosophised about nature. Other works attributed to Pythagoras (such as *On Education* and *On Statesmanship*) are discussed, as is the report, following an account given by Sosicrates (probably mid-second century BCE), that Pythagoras answered Leon the tyrant of Phlius' question about who he was with 'a philosopher' because, as Pythagoras went on to explain, the philosopher seeks truth (8.8). The section ends with a brief summary (8.9) of the contents of Pythagoras' three works, noting that he forbade prayer on the

---

[13] All translations of Diogenes Laertius in what follows are by Hicks 1925.

grounds that we do not know what will help us, called for temperance in drinking and eating, and offered cautions about the harmful effects of too much sexual activity.

A number of Pythagoras' accomplishments are then reported, in a seemingly random order. So, for example, we are told (8.11) that once, disrobed, his thigh appeared to be gold, and that as he was crossing the river Nessus (in Thrace[14]), a number of people heard the river welcome him. Then (8.11–12) we learn that he first led geometry to perfection, spending most of his time on the arithmetical aspect of geometry and also discovering the musical intervals on the monochord, while 'nor did he neglect even medicine'. Diogenes Laertius reports: 'We are told by Apollodorus the calculator that he offered a sacrifice of oxen on finding that in a right-angled triangle the square on the hypotenuse is equal to the [sum of the] squares on the sides containing the right angle' (1925: vol. 2, 331). Following the report of this discovery, we learn that Pythagoras was the first to recommend that athletes have a diet of meat rather than dried figs and soft cheese, but Diogenes Laertius notes that this may have been the work of a different Pythagoras, one who was an athletic trainer (8.12). In any case, Diogenes Laertius takes this opportunity to discuss Pythagoras' views on diet, including his vegetarianism (8.13). In this account, we find the mingling of details relating to wonders and marvels associated with Pythagoras, as well as accounts of his mathematical accomplishments. Diogenes Laertius specifically mentions some of his sources, and even questions some of the information he transmits.

We then learn something of Pythagoras' views on reincarnation, and that, according to Aristoxenus, he 'was the first to introduce weights and measures into Greece'; moreover, 'it was he who first declared that the Evening and Morning Stars are the same, as Parmenides maintains' (8.14). Immediately following this information, Diogenes Laertius reports that 'so greatly was he admired that his disciples used to be called "prophets to declare the voice of God", besides which he himself says in a written work that "after two hundred and seven years in Hades he has returned to the land of the living"' (1925: vol. 2, 333). Here we have an intellectual history, that is, a history of ideas and intellectual discoveries, intertwined with an aretalogical account, which (although it is not much developed) includes references to disciples as well as a resurrection story.[15]

---

[14] Cf. Hesiod *Theogony* 341. See also Smith 1867: 1169.
[15] Resurrection is not a standard feature of Diogenes Laertius' accounts of philosophers' lives. Accounts of discoveries – sometimes referred to as 'heurematology' or 'heurematography' – may constitute another genre. A useful starting point on the topic is Copenhaver 1978 and Zhmud 2006.

We are told (8.15) that 'down to the time of Philolaus [*c.* 470–390 BCE] it was not possible to acquire knowledge of any Pythagorean doctrine, and Philolaus alone brought out those three celebrated books which Plato sent a hundred minas to purchase' (1925: vol. 2, 335). We learn a little more about Pythagoras' teaching and lecturing and his views on education, and also about his gift for friendship. A number of his 'watchwords' or 'precepts' follow (at 8.17); Diogenes Laertius glosses the meaning of these sometimes cryptic sayings. He discusses Pythagoras' diet, appearance and habits (8.19–20), his dealings with oracles and priestesses (8.21) and his ethical advice to his disciples (8.22–24).

Diogenes Laertius then names Alexander (Polyhistor; first half of the first century BCE, who reported what he found in the Pythagorean memoirs or notebooks, *Pythagorika hupomnēmata*[16]) as his source for a rather lengthy list of Pythagoras' ideas on a range of topics. These topics are those which, traditionally, interest historians of science and mathematics: 'the principle of all things is the monad or unit'; 'the sun, the moon, and the other stars are gods; for, in them, there is a preponderance of heat, and heat is the cause of life'; 'the moon is illumined by the sun' (8. 25–27, trans. Hicks 1925: vol. 2, 341–343). This section is carefully demarcated by Diogenes Laertius, both at the beginning and at the end, as information coming from Alexander (8.24–36).

At 8.36 Diogenes Laertius notes that 'what follows is Aristotle's'. Diogenes repeatedly makes an effort to distinguish and name the source of his information; he then provides a series of anecdotes from a variety of sources (8.36–38), along with several different accounts of Pythagoras' death (8.39–40), an anecdote attributed to Hermippus (presumably of Smyrna, mid-third century BCE) regarding Pythagoras and his mother (regarding a journey into Hades),[17] and information about Pythagoras' wife Theano and his son Telauges (his successor, and possibly a teacher of Empedocles) (8.42–43). Diogenes Laertius (8.45) also offers information about his date of flourishing (60th Olympiad) and the longevity of his school (nine or ten generations). We also learn about other men of the same name living around the same time (8.46–48). A few brief tributes are then presented (8.48): 'Favorinus says that our philosopher used definitions throughout the subject matter of mathematics'; 'further, we are told he was the first to call the heaven the universe and the earth spherical'

---

[16] See Kahn 2001: 74–76 on this work.
[17] See Bar-Kochva 2010: 164–205 on Hermippus' biography of Pythagoras; Bar-Kochva argues that Hellenistic Jews had their own reasons for attributing to Pythagoras the transfer of Jewish customs and ideas to Greek philosophy.

(1925: vol. 2, 365). A letter from Pythagoras to Anaximenes is quoted (8:49–50), and Diogenes Laertius concludes his *bios* of Pythagoras.

Throughout the *Lives of Eminent Philosophers*, as in his 'life' of Pythagoras, Diogenes Laertius aimed to provide an account of intellectual lineage and relationships: the organisation of the work into 'schools' or 'successions' makes this plain. In addition, Diogenes Laertius' work is clearly a compilation, and he is proud to name his sources at various points. His account of the life of Pythagoras is not presented as a continuous narrative, but rather as a collection of information about his subject, who is only one among many in the *Lives of Eminent Philosophers*. But Pythagoras holds a special place in Diogenes Laertius' history of philosophy: he was the founder of a philosophical succession.

Porphyry's *Life of Pythagoras*

The status of the philosopher as hero, moral exemplar and even divinity is crucial in the 'biographies' discussed here. Scholars have pointed to the rising influence of Christianity as important in shaping the *bioi* of Pythagoras produced by both Porphyry and Iamblichus. Porphyry and his student Iamblichus were Neopythagoreans or Neoplatonists, depending on how one interprets these labels.[18] In their accounts, Pythagoras is, as Kahn (2001: 134) has noted, 'the paradigm of the sage as divine man'. Porphyry and Iamblichus emphasise the marvels and wonders associated with Pythagoras. Indeed, a number of twentieth-century scholars have pointed to parallels between the extant *Lives* of Pythagoras and the gospels (Lévy 1927; Dillon and Herschbell 1991); there are indications that both Porphyry and Iamblichus may have written their accounts with the specific goal of providing an alternative to the Christian gospels (Kahn 2001: 134; Dillon and Herschbell 1991: 25–26; see also O'Meara 1989: 214).

Porphyry's *Life of Pythagoras* was part of his *Philosophical History* from Homer to Plato; it is not complete, and our version ends abruptly. (Except for short extracts, it is the only part of the *Philosophical History* to survive.) Dominic O'Meara (1989: 26) has noted that Porphyry's *Life* 'reads like a learned compilation of source materials concerning Pythagoras'. His account shares some similarities of organisation with that of Diogenes

---

[18] See, e.g., Gorman 1979: 2; O'Meara 1989: 4–5 on the appropriation of Pythagorean ideas by Neoplatonic philosophers, and Kahn 2001: 133 on the absorption of neo-Pythagoreanism into neo-Platonism. Porphyry and Iamblichus shared a teacher-student relationship; Iamblichus is often described as a rival to Porphyry; see O'Meara 1989: 214.

Laertius, and some content, but contains more information and is more of a continuous narrative. Like Diogenes Laertius, Porphyry begins with a consideration of Pythagoras' origins and family (including his offspring), naming a number of sources that offer divergent accounts (1–5). But Porphyry, unlike Diogenes Laertius, does not begin his 'life' of Pythagoras by placing him within a particular intellectual tradition. In his opening passages, while commenting on Pythagoras' childhood precociousness, Porphyry draws attention to his family background and parentage: 'It is agreed by most that Pythagoras was the son of Mnesarchus, but there has been disagreement concerning Mnesarchus' race. Some say he was a Samian, but Neanthes in the fifth book of his work on Pythagoreanism says he was a Syrian'. He indicates that there are differing versions of Pythagoras' origins, and relates that:

> Apollonius, in his book *On Pythagoras*, also gives the name of Pythagoras' mother, Pythaïs, a descendant of Ancaeus, the founder of Samos. Apollonius says, too, that some declare Pythagoras was by procreation the child of Apollo and Pythaïs, and only nominally the child of Mnesarchus. As a matter of fact, one of the Samian poets does say, 'Pythagoras, the friend of Zeus, whom Pythaïs, most beautiful of the Samians, bore to Apollo.' Finally, that he was a pupil not only of Pherecydes and Hermodamas but of Anaximander too, is also stated by Apollonius.[19]

Following his recounting of the various accounts of Pythagoras' ancestry, Porphyry, like Diogenes Laertius, then moves on to the topic of Pythagoras' education (6–13), again naming his sources. Several anecdotes are offered (14–15), including one regarding the training of athletes and recommendations regarding diet (again, meat rather than cheese and figs).

Porphyry's order of exposition corresponds more closely to that described by Delatte than that of Diogenes Laertius. Porphyry recounts Pythagoras' travels once he left Samos for Italy, and the founding of his school (16–22). The description of Pythagoras' character is accomplished through illustrative anecdotes; Porphyry notes (28–29) that 'ten thousand other things yet more marvelous and more divine are told about the man, and told uniformly in stories that agree with each other'; indeed, he claims that 'to put it bluntly, about no one else have greater and more extraordinary things been believed' (1965: 116).

Information about his listening to the harmony of all things, including the harmony of the spheres, is reported (30–31), as are matters related to

---

[19] Trans. Smith (in Hadas and Smith) 1965: 107–108. All translations of Porphyry in what follows are by Smith, in that volume.

his daily life, including his diet and sacrificial practices (32–36). His teaching methods and ideas are discussed at some length (36–53), including their development within his school. Porphyry reports (46–47) that 'he practised a philosophy of which the object was to deliver and set free of such fetters and bonds [as incarnation] the mind that had been separated from the cosmic mind for incarnation in us'; 'accordingly, it uses mathematics and the sciences dealing with the borderland between bodies and the bodiless to train in advance the eyes of the soul' (1965: 122).

In contrast to Diogenes Laertius' presentation, Porphyry's account provides information about Pythagoras's ideas and teaching in a more unified manner; the references to Pythagoras' ideas and teaching are not scattered throughout the *bios* of Pythagoras, as they are in Diogenes Laertius' account. As noted earlier, Diogenes Laertius' ordering and changes in topic may reflect his own shifting between sources. Porphyry's account seems to be more digested, coherent and unified, and may allow readers to attend more readily to Pythagoras' intellectual achievements.

Iamblichus' *On the Pythagorean Life*

Iamblichus presents by far the lengthiest account of Pythagoras' life, but his work, *On the Pythagorean Life*, is arguably not or not only a biography, but rather a preliminary guide to Pythagorean philosophy. It was part of his larger work, *On Pythagoreanism*,[20] and is somewhat different from those presented by Diogenes Laertius and Porphyry. Kahn (2001: 5) credits Eduard Zeller (1880–1892) with being the first to point out that 'the further a document is from Pythagoras' own time, the fuller the account of Pythagoras becomes'; however, the length and detail of an account do not reflect its accuracy.

Iamblichus' account was more concerned with Pythagoras' divinity than the others were. It may be deliberately anti-Christian, and may also have been consciously modelled on and in competition with the Christian gospels (Clark 1989: ix–xiii; O'Meara 1989: 214–215). Furthermore, Iamblichus' account not only considers Pythagoras himself but also offers a good deal of information concerning his followers and the members of his 'school'. So, for example, *On the Pythagorean Life* ends with a list of the names of the 'most famous' Pythagorean men and women (including Spartans).

---

[20] See O'Meara 1989: 30–40 on the place of the *Life* within the larger work, and a consideration of its possible title.

From the very beginning of the work, Iamblichus invokes a completely different style from that of both Diogenes Laertius and Porphyry. He begins (chap. 1) with the explanation that:

> At the start of every philosophical investigation, it is after all the custom, at least for all who are sound-minded, to invoke God. But at the outset of that philosophy rightly believed to be named after the divine Pythagoras, it is surely all the more fitting to do this; for since this philosophy was at first handed down by the gods, it cannot be comprehended without the gods' aid. Moreover, its nobility and greatness exceed human ability to understand it immediately: only when the goodwill of the gods leads the way, can someone with gradual approach slowly appropriate something from it.[21]

Iamblichus makes it clear from the start that philosophy is not only an intellectual endeavour: it is dependent upon its link to the divine for its accomplishment. Indeed, Iamblichus proclaims that Pythagoras himself will be the divine guide through whom his philosophy can be comprehended: 'And after the gods, we shall choose as our leader the founder and father of this divine philosophy' (Dillon and Herschbell 1991: 31). In other words, the divine philosopher Pythagoras himself will be our teacher.

Iamblichus' account of Pythagoras' life and his school is a highly developed aretalogical account of an heroic, indeed divine, teacher; some might argue that it verges on the hagiographical. Iamblichus does mention, somewhat briefly, Pythagoras' ideas and 'discoveries' in *On the Pythagorean Life*, but the purpose of this first book of *On Pythagoreanism* was not to expound and explain Pythagorean philosophy in detail (O'Meara 1989: 33–34); this was the purpose of the later books, namely the *Protreptic to Philosophy* (Book 2), *On General Mathematical Science* (3), *On Nicomachus' Arithmetical Introduction* (4), *On Arithmetic in Physical Matters* (5), *On Arithmetic in Ethical Matters* (6), *On Arithmetic in Theological Matters* (7), *On Pythagorean Geometry* (8), *On Pythagorean Music* (9) and (probably) *On Pythagorean Astronomy* (Book 10). (Only the first four books of *On Pythagoreanism* survive; there are only excerpts of Books 5–7.)

Iamblichus' *On the Pythagorean Life* is itself an exhortation to adopt Pythagorean philosophy, illustrated and enhanced by the ideal model provided by the founder of that philosophy itself, namely Pythagoras himself (O'Meara 1989: 39). As Gillian Clark (1989: xv) has explained, once students of Pythagoreanism

> had read the life of Pythagoras, and become convinced that Pythagoras was a divine soul sent to reveal the truth and teach human beings how to live,

---

[21] Here and in what follows, translations of Iamblichus are by Dillon and Herschbell 1991; here, p. 31.

they were to continue with the *Protrepticus*, 'Exhortation to Philosophy', which offers Pythagorean sayings and philosophers side by side with extracts from Plato and Aristotle. Thus encouraged, they advanced to a series of highly technical works on aspects of Pythagorean mathematics: that is, mathematics understood as the study of the structure of reality.

In other words, *On the Pythagorean Life* provides the first step in one's indoctrination into Pythagoreanism.

It is only after the prologue describing the divinity of Pythagoras that Iamblichus begins his discussion of his origins (in chapter 2). Unlike Diogenes Laertius and Porphyry, Iamblichus starts his account of Pythagoras' life not by focusing on his hometown or biological parents, but instead describing Pythagoras' divine lineage: 'The story goes, then, that Ancaeus who dwelt in Same in Cephallenia was sired by Zeus'; 'the tradition is that Mnemarchus and Pythais, Pythagoras' parents, were from the household and family started by Ancaeus who founded the colony' (Dillon and Herschbell 1991: 33; on the variations of Pythagoras' father's name, see 35, n. 3). Pythagoras was celebrated as the descendant of a god, namely Zeus.

Iamblichus provides us with details of the way in which Pythagoras' father learned that his son would possess divine gifts and so renamed his wife and chose an appropriate name for his child, referring to the Pythian Apollo. He rejects suggestions that Apollo himself impregnated Pythagoras' mother. Nevertheless, Iamblichus assures us that 'no one would dispute, judging from his very birth and the all around wisdom of his life, that Pythagoras' soul was sent down to humans under Apollo's leadership, either as a follower in his train, or united with this god in a still more intimate way' (Dillon and Herschbell 1991: 35).

After the account of Pythagoras' divine lineage, or genesis, Iamblichus begins his description of his life, touching on his childhood, education and travels prior to his return to Samos at about age fifty-six (chapters 2–4). Following details of his life in Samos (5), we learn of his move to Italy and receive a brief description of his character and philosophy (6). Iamblichus is occasionally repetitive and sometimes contradictory; he tends not to name his sources, but in many instances offers a wealth of detail. So, for example, we learn (2) that Pythagoras travelled to see several of the pre-Socratic philosophers, and 'as he visited each in turn, the result of association with him was such that all cherished him and admired his character, and made him a partner in their discourses. And what is more, Thales gladly accepted him as a student, and admired his difference from other youths' (Dillon and Herschbell 1991: 39).

Iamblichus discusses Pythagoras' speeches in Italy (7), his visit to and teachings in Croton (8–11), devoting one chapter (10) to outlining Pythagoras' advice to children ('never start a fight') and another on his address to the women of Croton (11), in which he commended the feminine practice of lending clothes and jewellery when needed without requiring a witness. Following his explanation of why Pythagoras was the first person to call himself a philosopher (12), he offers examples of his ability to teach animals, and his views regarding transmigration of souls (13–14). The importance of music in education and restoring the soul (15), his views on friendship (16) and his relationships with his followers are described, and some of his maxims reported (17–18). Iamblichus (19) explained that Pythagoras 'discovered many ways of teaching and training, and transmitted the appropriate portion of wisdom according to each one's own nature and ability' (Dillon and Herschbell 1991: 115); the practices of Pythagorean philosophy are then described (20–24). Iamblichus returned to the topic of education through music (25) and Pythagoras' discovery of the principles of harmony (26). His political activities and the civic benefits bestowed by Pythagoras and his followers on humanity are described in some detail (27). Various marvels and miracles associated with Pythagoras, as well as his piety, are then celebrated (28); this is a particularly lengthy chapter. The wisdom of Pythagoras, including his work in physics, ethics and logic, as well as his study of geometry, are described briefly (29). His contributions to justice (30), his practice of self-control (31), his precepts on courage (32) and his teachings on friendship (33) are recounted, followed by customs of the Pythagoreans (34), the various political problems encountered by the Pythagoreans and the way the sect continued to operate (35). *On the Pythagorean Life* concludes with a list of his successors, including the seventeen 'most famous' Pythagorean women (chap. 36).[22] Pythagoras' students, followers and disciples are especially present, serving to emphasise his role as a teacher as well as the founder and leader of a community committed to a particular way of life. Iamblichus celebrated Pythagoras as an heroic teacher and divine guide to living.

But, of course, Pythagoras was not the only such guide to living available during the third century. Diogenes Laertius promoted Epicurus as a great ethical teacher. Varieties of Christianity were developed with the common aim of promoting Jesus. The Christian threat to pagan philosophers was certainly known to Porphyry and Iamblichus; Porphyry wrote a critique of Christianity, *Against the Christians*. In composing accounts

---

[22] See Clark 1989: xvi–xviii on the place of women within Pythagoreanism.

of the life of Pythagoras, our authors combined intellectual history with moral, ethical and, in some cases, apparently religious concerns. As Hadas explained, aretalogical accounts could be shaped by 'the level of the audience to which the aretalogy might be addressed, and the general religious climate that fostered its growth' (Hadas and Smith 1965: 62). Our ancient authors provided accounts of Pythagoras as a philosopher, but these accounts were shaped by various factors, including the authors' own, differing philosophical allegiances.[23]

Iamblichus' may well have been shaped by the Christian climate in which it was written. O'Meara has suggested that 'Iamblichus' successors could of course come to consider his Pythagorean theology as a revelation of much greater antiquity and purity than that of the Christians' (O'Meara 1989: 215). Nevertheless, as he has indicated (O'Meara 1989: 214),

> it would be difficult to show, on the basis of the extant remains of *On Pythagoreanism*, that Iamblichus had Christianity specifically in mind as a target against which his Pythagoreanizing programme was to be directed. At most, one could point to structural parallels between his figure of Pythagoras (his divine authority, attributes, mission, words and deeds among men) and Christ.

Looking at structure, John Dillon and Jackson Herschbell have argued that Iamblichus' text is not really a biography and should not be understood as an example of that genre. As they point out, 'much of the central part of the work seems hardly excusable in a biography: it concerns not Pythagoras himself, but the Pythagoreans in general'. They make the case that 'the work is, in fact, a dramatised study of a way of life, with a strong protreptic purpose', exhorting and urging the reader towards Pythagorean philosophy and the Pythagorean way of life. Dillon and Herschbell (1991: 25) go so far as to suggest that 'if it is permissible (that is, if it be correct to recognise it as a genre transcending the strictly Christian milieu), it seems best to classify [Iamblichus' work] as a gospel'.[24]

The survival of Iamblichus' *On the Pythagorean Life* and the loss of most of the books of *On Pythagoreanism* may skew our perception of the emphasis on Pythagorean ideas. The order of books in Iamblichus' work reflects what is meant to be a pedagogical progression from that which is more general to the more difficult, and that which can be understood as being specifically Pythagorean. This progression is designed to lead the

---

[23] While these must be noted, they cannot be treated fully here.
[24] Whereas some scholars have pointed to similarities to the synoptic gospels, Dillon and Herschbell 1991: 26 suggest that Iamblichus' account most closely resembles the non-synoptic gospel of John.

soul towards greater things. The student of Pythagoreanism will study, first, Pythagoras and his school, then general and Pythagorean philosophy, followed by Pythagorean mathematics. The programme for Pythagorean mathematics included: general mathematical science, including arithmetic itself; arithmetic in physics, in ethics and in theology; geometry; music and (probably) astronomy. Iamblichus' stress on the role of mathematics in the personal achievement of divinity distinguished his account of Pythagoras, his teachings and his school from that of others. For Iamblichus, the 'specifically Pythagorean revelation consists in the "most scientific", unerring forms of knowledge, those having to do with pure, immaterial unchanging realities, namely mathematics and the study of true being and of the divine' (O'Meara 1989: 33–35, 89). In his view, mathematics prepared the soul to attain wisdom. Yet, in his *On the Pythagorean Life*, the philosophical and mathematical ideas are only briefly mentioned; in this work it is the divine character of Pythagoras that is highlighted.

Within the 'aretalogical' literature, and specifically these *bioi* of Pythagoras, the desire to describe the philosopher's ideas and teachings interacted with the aim of promoting the divinity of the heroic teacher. Indeed, as Hadas noted, an account of 'the teacher for whom supernatural claims are made asks for something like religious conversion'. He suggested that the image of such teachers survives 'because those who cherish the legend constitute something like an organized cult, with leaders concerned for its propagation and its adaptation to new climates and new conditions'. The dual aims of institutionalising teachings while promoting divine status interact with each other; 'mystics are endowed with rational doctrine and rational thinkers with a mystique' (Hadas and Smith 1965: 62).

We may end by asking: What are these three *bioi* of Pythagoras about? Pythagorean philosophy has often been understood as emphasising the value of mathematics.[25] But the ancient *bioi* considered here cannot be regarded simply as biographies of an important figure in the history of science. Rather, the three ancient biographies of Pythagoras belong to a genre born in a specific time and place. These accounts are not biographies as we understand the term, even while recognising that biography is a diverse genre; nor are they the gospels of a failed religion. Their purpose

---

[25] Whitehead 1925: 53–54 pointed to 'old Pythagoras, from whom mathematics, and mathematical physics, took their rise'. He suggested that Pythagoras had endowed these fields 'with the luckiest of lucky guesses', but then asked: 'or, was it a flash of divine genius, penetrating to the inmost nature of things'.

was to provide a history of an intellectual tradition, relating the interactions of a teacher and his students, and also to celebrate the achievements of an heroic philosopher whose 'life' was meant to serve as a guide for others on how to live, how to benefit from philosophy and how to be more divine.

Conclusion

In an earlier volume in this series, James Clackson noted that 'choices made by ancient writers are rarely without a wider significance'. In concluding his *Language and Society in the Greek and Roman Worlds*, he argued that 'in our investigation of the written remains of antiquity, we must be sure that we pay as much attention to the language of our written sources as their authors did' (Clackson 2015: 171, 175). Clackson is well aware that texts are not the only source of historical evidence; in fact, in the passage quoted, he makes particular reference to the physical objects with which writing was associated, and the physical forms of written texts themselves. Similarly, our historical evidence of scientific ideas and practices is not restricted to texts.[1] Indeed, some of the texts studied in this volume, including Eratosthenes' *Letter to King Ptolemy*, make clear reference to instruments and tools used in scientific, mathematical and technical work. Nevertheless, most of our evidence relating to ancient Greek and Roman science is found in written texts. In this volume I have highlighted the significance of formats – or genres – used by ancient technical authors to convey their ideas and methods. This approach has been motivated by the desire to take seriously the choices available to those authors, and also by the conviction that important historical information – not least about the context in which the text was composed – is conveyed by the use of a particular genre.

The intention here has been to explore the variety of formats used by authors of ancient Greek and Roman works on scientific subjects, whilst considering the intellectual and wider cultural contexts in which these works were produced. Today, we take for granted the numerous formats deemed suitable for communicating scientific work, including specialist

---

[1] On the material culture related to scientific work in Greco-Roman antiquity, see, for example, Vitruvius *On architecture*, Book 9 chapter 8 (on various inventors of sundials); Evans 1999; Taub 2002.

Conclusion

journal articles and introductory teaching texts as well as pieces in the 'popular' press. Modern readers are often surprised that a variety of formats was used for Greek and Roman scientific and mathematical texts. How many would have guessed that the *Greek Anthology* of poetry contains mathematical problems? And, even those texts that are read and reread by many – such as Lucretius' *On the nature of things* – may be read by some only for 'scientific' content (excerpting the technical bits while ignoring the original poetic form) or only as poetry, bypassing the natural-philosophical detail.

It is imperative to recognise that very different sorts of texts may be concerned with scientific subjects: the identification of the characteristics of a scientific or mathematical text may not always be entirely straightforward. Indeed, as has been emphasised throughout this volume, the formal diversity of Greco-Roman texts dealing with scientific and mathematical subjects argues for a nuanced understanding of the place of scientific thinking in broader culture, reminding us that it is not always easy to identify 'science', nor is it a simple task to label particular texts as 'scientific'. As we have seen, generalisations may be of limited value. The aims of the author of a text and its intended function are often best understood by investigating the specific historical context in which the work was produced; this has been an important undercurrent throughout this volume. I have adopted a 'case studies' approach here, to try to limit some of the difficulties of generalisation.

As has been previously noted, the predilection of historians of science, philosophy and mathematics to focus on ideas has often tended to include the assumption that these ideas can be seamlessly extracted from a text and understood, with no consideration of the formal features of that text. I contend that our understanding of scientific and mathematical ideas is enhanced by engaging with the 'medium' which conveys the message, for the medium is also indicative of other aspects of the broader cultures in which it is used. Necessarily, this volume has concentrated on a relatively small number of formats, genres or types of text concerned with the physical world and mathematics. Yet, even while focusing on a limited number of forms, we have seen the persistence of some (including the popularity of poetry) as well as the creation and flourishing of 'new' genres, such as the encyclopaedia. Particular genres offer information about intellectual communities within the Greco-Roman world, including those concerned with mathematics and explaining the physical world. For example, letters often give specific evidence of the relationships between the author and intended readers, including patrons, followers and members of correspondence networks; as we have seen, the letter was a format favoured by some members of the Greek mathematical community.

There is not sufficient space available here to provide a full survey of the different genres of Greco-Roman scientific texts, including some which were particularly long-lived and are familiar even today, such as the introductory teaching text. It is the variety of forms used to communicate Greek and Roman scientific ideas and methods that has inspired this book, and there are numerous other genres and types of text that were used to communicate ideas about the physical world and mathematics in Greco-Roman antiquity which are also worthy of study.

For example, the lecture was an important extended prose format used for communicating scientific and mathematical ideas. We know from ancient sources that students of some of the ancient philosophers took notes and thereby recorded their oral teaching and lectures. Indeed, scholars have argued that much of what survives of Aristotle's writings are actually lectures, or notes for lectures. For example, the Greek title of the work credited to Aristotle and known to us as the *Physics* is *Physikē akroasis*; *akroasis* may be translated as 'lecture' or 'hearing', and the title thereby rendered as *Lecture Course on Nature* (see Taub 2008a: 18). Other ancient works dealing with scientific subjects, often described as treatises, may also have begun as lectures; one such example is Cleomedes' (*c.* 200 CE) work known as *The Heavens*. At several points Cleomedes' language – referring to 'lecture courses' (*skholai*) – suggests that his work had its beginning as a series of lectures.[2] Here we have a glimpse of a recurring feature of Greco-Roman scientific discourse, which has been a sub-theme throughout this volume: the interplay and fluid relationships between oral and written presentations, attested in poetry (including the mathematical poems that may have been recited at symposia), as well as letters (which often begin conversationally) and even commentaries (sometimes reflecting a group discussion).

Yet, in spite of the recurring marks of oral culture pervasive in Greco-Roman scientific writings, here the focus has been on the consideration of these writings as texts. I have been primarily interested in what might be termed 'authorial choices', working with the recognition that – to some extent – ancient authors writing about scientific and mathematical topics had a range of formats and genres to choose from. I deliberately chose not to try to engage with issues relating to readership, while recognising

---

[2] Cleomedes 2.2.7, 2.7.12, trans. Bowen and Todd 2004: 127, 165. See also Taub 2008a: 19.

that the authorial choice of genre would have elicited various responses from potential readers. For example, some readers may have resisted certain texts because of the genre in which they were presented, while others may have been attracted to a specific format. Indeed, we have strong clues that certain genres would have had special appeal to readers, and that this would have been important to authors: Lucretius' view that the 'honeyed-cup' of poetry would attract more readers than would prose has already been mentioned earlier in this book.

Of course, Greco-Roman scientific writings were read not only in antiquity, but in subsequent periods, across various cultures, in their original languages and also in translation. Many of these texts had a very long 'afterlife' beyond their original readership, serving as exemplars of what scientific, mathematical and technical texts could – even should – look like, well in to the modern period, including that of the 'scientific revolution'. The genres used by Greek and Roman authors to communicate scientific material persisted, even when, at times, the original ideas conveyed were actively rejected.

One of the ambitions here has been to encourage more reading today of ancient scientific and technical texts, and more work to be done on texts as texts, particularly those that have not been much studied. We should study examples of ancient scientific, mathematical and technical writing because of the complexity they offer, reminding ourselves that the historical and cultural contexts of scientific, mathematical and technical discourse provide many layers and levels of meaning. Through the study of these writings as texts, we can see traces of the writers and communities of readers of scientific work, as they communicated ideas and practices embedded in wider culture. And there are other questions to be asked, including how metaphor fits within scientific discourse, and what sort of explanation can be conveyed via verse.

In the closing Bibliographical Essay, I provide additional suggestions for readers wishing to explore further texts in which scientific and mathematical ideas were presented in ancient Greece and Rome. The Bibliographical Essay also offers a preliminary road map to more traditional accounts of Greek and Roman science.

Throughout this volume, I have argued that the genres in which Greek and Roman scientific, mathematical and technical ideas were conveyed are worthy of study. Indeed, genre is one of the important bridges connecting authors and readers, for both authors and readers bring to texts expectations and shared tacit knowledge regarding specific genres of communication. It is, primarily, only through surviving texts that we have access to the scientific ideas of antiquity. Yet, by concentrating solely on the ideas conveyed in those texts, we may limit our understanding of the meaning of those ideas within their historical context. By considering the diverse ways in which scientific, mathematical and technical ideas were communicated, through different types of texts, we can uncover otherwise hidden meanings and more fully comprehend the historical contexts in which those ideas were produced and shared. What has been presented in this volume is intended to emphasise the value of reading these texts, as texts.

APPENDIX A

Arithmetical Epigrams from Book 14 of The Greek Anthology

Translated into prose by W.R. Paton (Cambridge, MA: 1918)

(with his solutions included)

116

Mother, why dost thou pursue me with blows on account of the walnuts? Pretty girls divided them all among themselves. For Melission took two-sevenths of them from me, and Titane took the twelfth. Playful Astyoche and Philinna have the sixth and third. Thetis seized and carried off twenty, and Thisbe twelve, and look there at Glauce smiling sweetly with eleven in her hand. This one nut is all that is left to me.

Solution: There were 336 (96 + 28 + 56 + 112 + 20 + 12 + 11 + 1).

117

A. Where are thy apples gone, my child? *B*. Ino has two-sixths and Semele one eighth, and Autonoe went off with one-fourth, while Agave snatched from my bosom and carried away a fifth. For thee ten apples are left, but I, yes I swear it by dear Cypris, have only this one.

Solution: There were 120 (40 + 15 + 30 + 24 + 11).

118

Myrto once picked apples and divided them among her friends; she gave the fifth part to Chrysis, the fourth to Hero, the nineteenth to Psamathe, and the tenth to Cleopatra, but she presented the twentieth part to Parthenope and gave only twelve to Evadne. Of the whole number a hundred and twenty fell to herself.

Solution: 380 (76 + 95 + 20 + 38 + 19 + 12 + 120).

119

Ino and Semele once divided apples among twelve girl friends who begged for them. Semele gave them each an even number and her sister an odd number, but the latter had more apples. Ino gave to three of her friends three-sevenths, and to two of them one-fifth of the whole number. Astynome took eleven away from her and left her only two apples to take to the sisters. Semele gave two quarters of the apples to four girls, and to the fifth one sixth part, to Eurychore she made a gift of four; she remained herself rejoicing in the possession of the four other apples.

Solution: Ino distributed 35 (15 + 7 + 11 + 2) and Semele 24 (12 + 4 + 4 + 4).

120

The walnut-tree was loaded with many nuts, but now someone has suddenly stripped it. But what does he say? "Parthenopea had from me the fifth part of the nuts, to Philinna fell the eighth part, Aganippe had the fourth, and Orithyia rejoices in the seventh, while Eurynome plucked the sixth part of the nuts. The three Graces divided a hundred and six, and the Muses got nine times nine from me. The remaining seven you will find still attached to the farthest branches."

Solution: There were 1,680 nuts.

121

From Cadiz to the city of the seven hills the sixth of the road is to the banks of Baetis, loud with the lowing of herds, and hence a fifth to the Phocian soil of Pylades—the land is Vaccaean, its name derived from the abundance of cows. Thence to the precipitous Pyrenees is one-eighth and the twelfth part of one-tenth. Between the Pyrenees and the lofty Alps lies one-fourth of the road. Now begins Italy and straight after one-twelfth appears the amber of the Po. O blessed am I who have accomplished two thousand and five hundred stades journeying from thence! For the Palace on the Tarpeian rock is my journey's object.

Solution: The total distance is 15,000 stades (say 1,500 miles); from Cadiz to the Guadalquivir, *i.e.* to its upper waters, 2,500, thence to the Vaccaei (south of the Ebro) 3,000, thence to the Pyrenees 2,000, thence to the Alps 3,750, thence to the Po 1,250, thence to Rome 2,500.

122

After staining the holy chaplet of fair-eyed Justice that I might see thee, all-subduing gold, grow so much, I have nothing; for I gave forty talents under evil auspices to my friends in vain, while, O ye varied mischances of men, I see my enemy in possession of the half, the third, and the eighth of my fortune.

Solution: 960 talents (480 + 320 + 120 + 40).

123

Take, my son, the fifth part of my inheritance, and thou, wife, receive the twelfth; and ye four sons of my departed son and my two brothers, and thou my grieving mother, take each an eleventh part of the property. But ye, my cousins, receive twelve talents, and let my friend Eubulus have five talents. To my most faithful servants I give their freedom and these recompenses in payment of their service. Let them receive as follows. Let Onesimus have twenty-five minae and Davus twenty minae, Syrus fifty, Synete ten and Tibius eight, and I give seven minae to the son of Syrus, Synetus. Spend thirty talents on adorning my tomb and sacrifice to Infernal Zeus. From two talents let the expense be met of my funeral pyre, the funeral cakes, and grave-clothes, and from two let my corpse receive a gift.

Solution: The whole sum is 660 talents (132 + 55 + 420 + 12 + 5 + 2 + 34).

124

The sun, the moon, and the planets of the revolving zodiac spun such a nativity for thee; for a sixth part of thy life to remain an orphan with thy dear mother, for an eighth part to perform forced labour for thy enemies. For a third part the gods shall grant thee home-coming, and likewise a wife and a late-born son by her. Then thy son and wife shall perish by the spears of the Scythians, and then having shed tears for them thou shalt reach the end of thy life in twenty-seven years.

Solution: He lived 72 years (12 + 9 + 24 + 27).

125

I am a tomb and I cover the lamented children of Philinna, containing fruit of her vainly travailing womb such as I describe. Philinna gave me my fifth portion of young men, my third of maidens, and three newly married daughters; the other four descended to Hades from her womb without participating at all in the sunlight and in speech.

Solution: She had 15 children (3 + 5 + 3 + 4).

126

This tomb holds Diophantus. Ah, how great a marvel! The tomb tells scientifically the measure of his life. God granted him to be a boy for the sixth part of his life, and adding a twelfth part to this, he clothed his cheeks with down; He lit him the light of wedlock after a seventh part, and five years after his marriage He granted him a son. Alas! late-born wretched child; after attaining the measure of half his father's life, chill Fate took him. After consoling his grief by this science of numbers for four years he ended his life.

Solution: He was a boy for 14 years, a youth for 7, at 33 he married, at 38 he had a son born to him who died at the age of 42. The father survived him for 4 years, dying at the age of 84.

127

Demochares lived for a quarter of his whole life as a boy, for a fifth part of it as a young man, and for a third as a man, and when he reached grey old age he lived thirteen years more on the threshold of eld.

Solution: He lived 15 years as a boy, 12 as a young man, 20 as a man, and 13 years as an old man; in all 60.

128

What violence my brother has done me, dividing our father's fortune of five talents unjustly! Poor tearful I have this fifth part of the seven-elevenths of my brother's share. Zeus, thou sleepest sound.

Solution: The one offered is that the one brother had $4^4/_{11}$ of a talent, the other $7/_{11}$, but I cannot work it out.

129

A traveller, ploughing with his ship the broad gulf of the Adriatic, said to the captain, "How much sea have we still to traverse?" And he answered him, "Voyager, between Cretan Ram's Head and Sicilian Peloris are six thousand stades, and twice two-fifths of the distance we have traversed remains till the Sicilian strait."

Solution: They had travelled 3,333⅓ stades and had still 2,666⅔ to travel.

130

Of the four spouts one filled the whole tank in a day, the second in two days, the third in three days, and the fourth in four days. What time will all four take to fill it?

Answer: ¹²/₂₅ of a day.

131

Open me and I, a spout with abundant flow, will fill the present cistern in four hours; the one on my right requires four more hours to fill it, and the third twice as much. But if you bid them both join me in pouring forth a stream of water, we will fill it in a small part of the day.

Answer: In 2²/₁₁ hours.

132

This is Polyphemus the brazen Cyclops, and as if on him someone made an eye, a mouth, and a hand, connecting them with pipes. He looks quite as if he were dripping water and seems also to be spouting it from his mouth. None of the spouts are irregular; that from his hand when running will fill the cistern in three days only, that from his eye in one day, and his mouth in two-fifths of a day. Who will tell me the time it takes when all three are running?

Answer: ⁶/₂₃ of a day.

133

What a fine stream do these two river-gods and beautiful Bacchus pour into the bowl. The current of the streams of all is not the same. Nile flowing alone will fill it up in a day, so much water does he spout from his paps, and the thyrsus of Bacchus, sending forth wine, will fill it in three days, and thy horn, Achelous, in two days. Now run all together and you will fill it in a few hours.

Answer. $6/11$ of a day.

134

O woman, how hast thou forgotten Poverty? But she presses hard on thee, goading three ever by force to labour. Thou didst use to spin a mina's weight of wool in a day, but thy eldest daughter spun a mina and one-third of thread, while thy younger daughter contributed a half-mina's weight. Now thou providest them all with supper, weighing out one mina only of wool.

Answer. The mother in a day $6/17$, the daughters, respectively, $8/17$ and $3/17$.

135

We three Loves stand here pouring out water for the bath, sending streams into the fair-flowing tank. I on the right, from my long-winged feet, fill it full in the sixth part of a day; I on the left, from my jar, fill it in four hours; and I in the middle, from my bow, in just half a day. Tell me in what a short time we should fill it, pouring water from wings, bow, and jar all at once.

Answer. $1/11$ of a day.

136

Brick-makers, I am in a great hurry to erect this house. Today is cloudless, and I do not require many more bricks, but I have all I want but three hundred. Thou alone in one day couldst make as many, but thy son left off working when he had finished two hundred, and thy son-in-law when he had made two hundred and fifty. Working all together, in how many hours can you make these?

Answer. $2/5$ of a day.

137

Let fall a tear as you pass by; for we are those guests of Antiochus whom his house slew when it fell, and God gave us in equal shares this place for a banquet and a tomb. Four of us from Tegea lie here, twelve from Messene, five from Argos, and half of the banqueters were from Sparta, and Antiochus himself. A fifth of the fifth part of those who perished were from Athens, and do thou, Corinth, weep for Hylas alone.

Solution: There were 50 guests.

138

Nicarete, playing with five companions of her own age, gave a third of the nuts she had to Cleis, the quarter to Sappho, and the fifth to Aristodice, the twentieth and again the twelfth to Theano, and the twenty-fourth to Philinnis. Fifty nuts were left for Nicarete herself.

Solution: She had 1,200 nuts (400 + 300 + 240 + 160 + 50 + 50).

139

Diodorus, great glory of dial-makers, tell me the hour since when the golden wheels of the sun leapt up from the east to the pole. Four times three-fifths of the distance he has traversed remain until he sinks to the western sea.

Answer: 3 hours and $9/17$ had passed, 8 hours and $8/17$ remained.

140

Blessed Zeus, are these deeds pleasing in thy sight that the Thessalian women do in play? The eye of the moon is blighted by mortals; I saw it myself. The night still wanted till morning twice two-sixths and twice one-seventh of what was past.

Solution: $6^6/_{41}$ of the night had gone by and $5^{35}/_{41}$ remained.

141

Tell me the transits of the fixed stars and planets when my wife gave birth to a child yesterday. It was day, and till the sun set in the western sea it wanted six times two-sevenths of the time since dawn.

Answer: It was $4^8/_{19}$ hours from sunrise.

142

Arise, work-women, it is past dawn; a fifth part of three-eighths of what remains is gone by.

Answer: $36/43$ of an hour had gone by.

143

The father perished in the shoals of the Syrtis, and this, the eldest of the brothers, came back from that voyage with five talents. To me he gave twice two-thirds of his share, on our mother he bestowed two-eighths of my share, nor did he sin against divine justice.

Solution: The elder brother had $1^5/_7$ talents, the younger $2^2/_7$, the mother 1 talent.

144

A. How heavy is the base I stand on together with myself! *B*. And my base together with myself weighs the same number of talents. *A*. But I alone weigh twice as much as your base. *B*. And I alone weigh three times the weight of yours.

Answer: From these data not the actual weights but the proportions alone can be determined. The statue A was a third part heavier than B, and B only weighed ¾ of the statue A. The base of B weighed thrice as much as the base of A.

145

A. Give me ten minas and I become three times as much as you. *B*. And if I get the same from you I am five times as much as you.

Answer: $A = 15^5/_7$, $B = 18^4/_7$.

146

A. Give me two minas and I become twice as much as you. *B*. And if I got the same from you I am four times as much as you.

Answer: $A = 3^5/_7$, $B = 4^6/_7$.

147

Answer of Homer to Hesiod when he asked the Number of the Greeks who took part in the War against Troy. There were seven hearths of fierce fire, and in each were fifty spits and fifty joints on them. About each joint were nine hundred Achaeans.

Answer: 315,000.

APPENDIX B

Eratosthenes' Letter to King Ptolemy

Trans. I.E. Drabkin (in Cohen and Drabkin) 1948: 62–66.

Reproduced from *A Source Book in Greek Science* by Morris R. Cohen and I. E. Drabkin, Cambridge, MA: Harvard University Press. Copyright © 1948 by the President and Fellows of Harvard College.

Eratosthenes to King Ptolemy, greetings.

The story goes that one of the ancient tragic poets represented Minos having a tomb built for Glaucus, and that when Minos found that the tomb measured a hundred feet on every side, he said: "Too small is the tomb you have marked out as the royal resting place. Let it be twice as large. Without spoiling the form quickly double each side of the tomb."

This was clearly a mistake. For if the sides are doubled the surface is multiplied fourfold and the volume eightfold.

Now geometers, too, sought to find a way to double the given solid without altering its form. This problem came to be known as the duplication of the cube, for, given a cube, they sought to double it. Now when all had sought in vain for a long time, Hippocrates of Chios first discovered that if a way can be found to construct two mean proportionals in continued proportion between two given straight lines, the greater of which is double the lesser, the cube will be doubled. So that his difficulty was resolved into another no less perplexing.

Some time later certain Delians, they say, seeking by order of the oracle to double an altar, fell into the same difficulty. And so they sent representatives to ask the geometers of Plato's school in the Academy to find the solution for them. These geometers zealously tackled the problem of finding two mean proportionals between two given lines. And Archytas of Tarentum is said to have obtained a solution with semicylinders, while Eudoxus used so-called 'curved lines'. All who solved the problem succeeded in finding the deductive proof, but they were not able to demonstrate the construction in a practical and useful way, with the exception

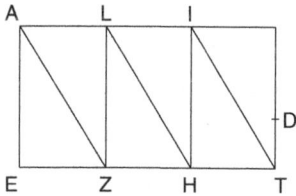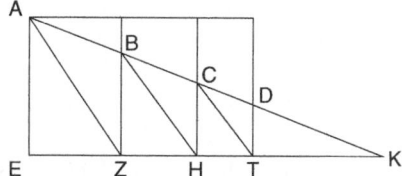

Figure A.1 Diagrams based on Cohen and Drabkin 1948: 64, to accompany Eratosthenes' *Letter to King Ptolemy*.

of Menaechmus (though he accomplished this only to a very small degree and with difficulty).

Now I have discovered an easy method of finding, by the use of an instrument, not only two but as many mean proportionals as desired between two given lines. With this discovery we shall be able to convert into a cube any given solid whose surfaces are parallelograms, or to change it from one form to another, and, again, to construct a solid of the same form as the given solid but larger, i.e., preserving the similarity. And we shall also be able to apply this in constructing altars and temples. We shall be able, furthermore, to convert our liquid and dry measures, the metretes and the medimnus, into a cube, and from the side of this cube to measure the capacity of other vessels in terms of these measures. My method will also be useful for those who wish to increase the size of catapults and ballistas. For, if the throw is to be increased, all the elements of these engines, the thicknesses, lengths, and the sizes of the openings, wheel casings, and cables must be increased in proportion. But this cannot be done without finding the mean proportionals. I have described below for you the demonstration and the method of construction of my device.

Let two unequal straight lines, AE and DT, be given, between which it is required to find two mean proportionals in continued proportion. Let AE be perpendicular to a line ET. Erect upon ET three [equal] parallelograms, AZ, ZI, and IT, in order. Draw diagonals AZ, LH, and IT. These diagonals will be parallel. Now while the middle parallelogram ZI remains fixed, let parallelograms AZ and IT be pushed so that AZ moves above and IT below the middle parallelogram, as in the second figure [shown], until A, B, C, and D lie along the same straight line. Draw this line and let it intersect ET, produced, at K.

AK : KB = EK : KZ (since AE and BZ are parallel),
and AK : KB = ZK : KH (since AZ and BH are parallel).

∴ AK : KB = EK : KZ = KZ : KH.
Again, since BK : KC = ZK : KH (since BZ and CH are parallel),
   and BK : KC = KH : KT (since BH and CT are parallel),
∴ BK : KC = ZK : KH = KH : KT.
But EK : KZ = ZK : KH.
∴ EK : KZ = ZK : KH = HK : KT.
But EK : KZ = AE : BZ,
and ZK : KH = BZ : CH,
and HK : KT = CH : DT.
∴ AE : BZ = BZ : CH = CH : DT.

That is, two mean proportionals, BZ and CH, have been found between AE and DT.

Such, then, is the demonstration on geometrical surfaces. But to find the two mean proportionals by an instrument, construct a frame of wood, ivory, or bronze having three equal flat surfaces as thin as possible. Let the middle surface be fixed, and the other two move along grooves; the size and shape of the surfaces may vary as desired. For the proof is not affected.

In order that the required lines may be obtained more accurately, care must be taken that when the surfaces are brought together all parts remain parallel, and fit one another snugly without gaps.

The instrument in bronze is placed on the votive monument beneath the crown of the column and is held fast with lead. Under the instrument the proof is set down concisely with a diagram and after this an epigram. These have been copied below for you, so that you may have them just as they are on the votive column. Of the two figures the second is engraved on the column.

'Given two straight lines, to find two mean proportionals in continued proportion. Let AE and DT be given. I draw together the surfaces in the instrument until points A, B, C, and D are all in a straight line. Consider these points as they are in the second diagram.

AK : KB = EK : KZ (since AE and BZ are parallel),
and AK : KB = ZK : KH (since AZ and BH are parallel).
∴ EK : KZ = KZ : KH = AE : BZ = BZ : CH.
Similarly we shall be able to show that
ZB : CH = CH : DT.
∴ AE, BZ, CH, and DT are in continued proportion.

That is, two mean proportionals between the two given lines have been found.

Now if the given lines are not equal to AE and DT, by taking AE and DT proportional to the given lines we shall obtain the means between AE and DT and then transfer the results to the given lines. Thus we shall have done what was required. And if it is required to find more mean proportionals, we shall achieve our purpose in each case by constructing one more surface on the instrument than the number of means required. The proof is the same in this case.

If, my friend, you seek to make from a small cube a cube twice as large, and readily convert any solid form into another, here is your instrument. You can, then, measure a fold, or a grain pit, or the broad hollow of a well, if between two rulers you find means the extreme ends of which converge. Do not seek the cumbersome procedure with Archytas' cylinders, or to make the three Menaechmian sections of the cone; seek not the type of curved line described by god-fearing Eudoxus. For with these plates of mine you could readily construct ten thousand means beginning with a small base.

You are a happy father, Ptolemy, because you enjoy youth with your son and have yourself given him all that is precious to the Muses and to kings. May he hereafter, heavenly Zeus, receive the scepter from your hand: so may it come to pass. And let whoever sees this votive column say: "This is an offering of Eratosthenes of Cyrene"'.

Bibliographical Essay

This bibliographical essay is intended to offer an introduction to ancient Greek and Roman science and to provide suggestions for further reading on the topics already discussed. Some works that are specifically relevant to textual studies are listed, with the hope that readers will be inspired to delve further into questions about the forms through which scientific and mathematical ideas were conveyed. Readers are also pointed towards other perspectives, including those of more traditional accounts of Greek and Roman science. Perhaps unsurprisingly, histories of science have often been informed by trends in other areas of historical inquiry. And scholars specialising in ancient philosophy also study ancient natural philosophy and mathematics, usually from a philosophical standpoint.

What follows is not an exhaustive reading list, but rather one which is meant to indicate some of those trends and to provide several possible avenues for engaging with histories of ancient science. First, surveys of ancient Greek and Roman science are noted, followed by accounts of the work of specific individuals, treatments focusing on particular subjects (such as physics and mathematics), and finally studies concentrating on textual aspects (including the genres) of Greek and Roman scientific and technical writings. The numerous 'handbooks' and 'companions' to ancient history, philosophy and culture (including science) also have useful bibliographies; several of these are mentioned here, as a sample of what is available.

(General) Histories of Ancient Science

For many readers, the works of Lloyd will be the starting point for the study of Greek Science, including his *Early Greek Science* (1970); *Methods and Problems in Greek Science* (1991); *Greek Science after Aristotle* (1973); *Magic, Reason and Experience: Studies in the Origins and Development of Greek Science* (1979), as well as the multiplicity of other volumes he has published. Lloyd's work is particularly informed by his close engagement with the history of philosophy, mathematics and medicine, as well as modern anthropological studies. Clagett's *Greek Science in Antiquity* (1957) does not restrict itself, as the title suggests, to Greek science, and this is one of this classic's many virtues: Clagett discusses Roman science and extends his account well into Late Antiquity, providing a detailed single-volume overview of Greek and Roman science. The work of Farrington, including *Greek Science* (1944/1961) and *Science and Politics in the Ancient World* (1939), has attracted generations of readers, even though some have distanced themselves from his Marxist and materialist interpretations. The Classical Association published Rihll's *Greek Science* (1999), which presents an overview organised by subject (physics, mathematics, astronomy, geography, biology and medicine) and reading lists. Other sorts of overviews are provided by edited volumes, such as Rihll and Tuplin (eds), *Science and Mathematics in Ancient Greek Culture* (2002) and *The Cambridge History of Science*, vol. 1, Jones and Taub (eds) (forthcoming). Reference works such as *The Oxford Classical Dictionary* (4th edn, 2012; Hornblower, Spawforth, and Eidinow (eds)), *A Companion to Ancient Philosophy*, Gill and Pellegrin (eds) (2006) and *The Oxford Handbook of Hellenic Studies* (2009, Boys-Stones, Graziosi, and Vasunia (eds)) have many valuable articles under various relevant headings. Even in antiquity, there was interest in histories of scientific inquiry; see Zhmud (2006), *The Origin of the History of Science in Classical Antiquity*.

The 'Great Man' Approach to the History of Science and Studies of Significant Individuals

Many histories of science and mathematics have centred on 'great' individuals deemed responsible for 'great' ideas – those regarded as having been most significant and influential over time. Amongst these figures of the ancient world, names such as those of Euclid (date uncertain, between 325 and 250 BCE), Aristotle (384–322 BCE), Archimedes (c. 287–212 BCE) and Galen of Pergamum (129–216 CE) loom large. Euclid is seen as responsible for setting the foundations of geometry, so important that the development of a different sort of geometry in the nineteenth century is referred to as 'non-Euclidean' geometry. Aristotle is generally regarded as one of the greatest scientists ever to have lived, and one of the only ancient philosophers to have focused so carefully on the study of animal life. The very high standard of mathematical originality and achievement set by Archimedes, amongst his other accomplishments, is still celebrated in the modern world; the Fields Medal (the International Medal for Outstanding Discoveries in Mathematics), established in the twentieth century, bears a portrait of him. And the ideas of the physician and philosopher Galen held sway over Western medicine for well over a thousand years. In the case of some renowned figures, such as Claudius Ptolemy, we know very little of their lives. Our knowledge of the achievements comes to us through ancient writings attributed to them, and through the written testimonies of others.

Strikingly, even in antiquity it was not entirely certain whether Thales of Miletus, often credited with having been the first philosopher (that is, lover of wisdom), had written anything (Diogenes Laertius 1.23); we know about his ideas through the accounts of others. Today, historians of mathematics understand that the work known as the *Elements*, traditionally attributed to Euclid, was actually the product of a number of people, a compilation rather than a single-authored work (Fowler 1999). The Aristotelian Corpus, the collection of writings associated with Aristotle, is generally understood to include a number of spurious works, which were probably not written by Aristotle himself (for example, *De mundo*, on cosmology). Indeed, many of the works in the corpus are thought to be 'lecture notes' rather than polished pieces of writing intended for wider circulation (see Introduction to this volume; Taub 2008a); Aristotle's 'authorship' of these works is not clear-cut. As we have seen in Chapter 1, the great mathematician Archimedes was credited not only with important mathematical works structured according to the Greek geometrical tradition, but also with the composition of an elegant epigrammatic poem, setting out an extraordinarily difficult mathematical problem; this foregrounds the 'writerly' activities and concerns of technical authors. In spite of the ancient attributions of the *Cattle Problem* to Archimedes, we cannot be certain of his role in its composition. Assertions of authorship were not always reliable: Galen complained that works not actually written by him were being circulated with his name attached (see his *My Own Books* 8–9; trans. Singer, 2002: 3).

It is those written texts that survive which serve as our principal source of information about science and mathematics in ancient Greece and Rome. Fortunately, in many cases we have good editions of these writings which are fairly easily accessible, for example, often (but not always) in the Loeb Classical Library series, now available online (www.hup.harvard.edu/features/loeb/digital.html); other valuable digital resources include the *Thesaurus Linguae Graecae* (http://stephanus.tlg.uci.edu/, containing digitised versions of most Greek literary texts from Homer to the fall of Byzantium in 1453, including many philosophical and technical texts) and the Perseus Digital Library (www.perseus.tufts.edu, available freely without subscription charge). Collections of shorter selections from primary sources in translation include: Cohen and Drabkin, *A Source Book in Greek Science* (1948), Irby-Massie and Keyser, *Greek Science of the Hellenistic Era: A Sourcebook* (2002), and Humphrey, Oleson and Sherwood, *Greek and Roman Technology: A Sourcebook* (2002).

As is to be expected, the literature on the great figures of ancient science – not only those mentioned earlier – is voluminous. Compiling a comprehensive bibliography is not the goal here. Nevertheless, I recognise that some readers will wish to know, for example, what Euclid did for mathematics, Aristotle for the study of living things, Ptolemy for astronomy and geography and Galen for medicine, and so I take this opportunity to suggest some books which I have found to be particularly engaging. Readers may also wish to consult articles on individuals in the *Dictionary of Scientific Biography (DSB)*, *The Oxford Classical Dictionary* (*OCD*) and Keyser and Irby-Massie (eds), *The Encyclopedia of Ancient Natural Scientists* (2008).

Studies Focusing on Specific Individuals

Thales of Miletus (*fl.* 585 BCE) is traditionally regarded as the first philosopher in ancient Greece, the first to explain phenomena 'naturally' without recourse to the traditional gods; he was also credited in antiquity with the ability to predict eclipses successfully (Herodotus 1. 74. 2). For those interested in Thales and other Presocratic philosophers, Kirk, Raven and Schofield (KRS), *The Presocratic Philosophers*, 2nd edn (1983) remains an excellent starting point, for the fragments as well as commentary; see also *The Texts of Early Greek Philosophy*, trans. and ed. Graham (2010). Pythagoras' life and work (as we saw in Chapter 5) have been the subject of various accounts beginning in antiquity; see Kahn, *Pythagoras and the Pythagoreans* (2001).

Some of the greatest ancient philosophers were distinguished by their contributions to natural philosophy and their views on mathematics. The literature on Plato is vast. One might begin by reading his *Timaeus*, and Vlastos, *Plato's Universe* (1975); Johansen, *Plato's Natural Philosophy* (2004); Broadie, *Nature and Divinity in Plato's Timaeus* (2012). Similarly, an enormous amount of work has been done on Aristotle. Readers will find the following (and the bibliographies contained therein) to be good starting points: Barnes, *Aristotle: A Very Short Introduction* (2000); Falcon, *Aristotle and the Science of Nature* (2005); Lloyd, *Aristotelian Explorations* (1998); Lennox, *Aristotle's Philosophy of Biology* (2001); Solmsen, *Aristotle's System of the Physical World* (1960); Judson, *Aristotle's Physics* (1991). Aristotle's student, colleague and successor as head of the Lyceum, Theophrastus, has been at the center of a series of studies produced by the Project Theophrastus of Rutgers University; a starting point is *Theophrastus of Eresus: On His Life and Work*, ed. Fortenbaugh, Huby and Long (1985). On Epicurus, see Asmis, *Epicurus' Scientific Method* (1984); on Lucretius, Sedley, *Lucretius and the Transformation of Greek Wisdom* (1998); see also Clay, *Lucretius and Epicurus* (1983).

Turning to great names in the history of mathematics, see the introduction, translation and commentary by Heath, *The Thirteen Books of Euclid's Elements* (1956); see also Fowler (1999) on Euclid's *Elements*. On Aristarchus, see Heath, *Aristarchus of Samos* (1981). On Archimedes: Netz and Noel, *The Archimedes Codex* (2007), is an heroic account with the flavour of a detective story, offering an appreciation of Archimedes' accomplishments; see also Jaeger, *Archimedes and the Roman Imagination* (2008). On Ptolemy: Graßhoff, *The History of Ptolemy's Star Catalogue* (1990); Taub, *Ptolemy's Universe* (1993); Berggren and Jones, *Ptolemy's Geography* (2000). On the 'greatest' physicians of antiquity, see Jouanna, *Hippocrates*, trans. DeBevoise (1999); Mattern, *The Prince of Medicine: Galen in the Roman Empire* (2013); and Gill, Whitmarsh and Wilkins (eds), *Galen and the World of Knowledge* (2009) for a collection of excellent essays discussing various aspects of Galen's work.

On one of the few known and renowned women of ancient mathematics and philosophy, see Dzielska, *Hypatia of Alexandria* (1995), regarded as a heroic figure for several reasons, including her death as a pagan martyr at the hands of Christians in 415. Another special hero – although not a natural philosopher or mathematician himself – is nevertheless sometimes portrayed as having died whilst investigating a natural phenomenon (the eruption of Vesuvius in 79 CE); Pliny the Elder was an important figure historically, whose *Natural History* discussed a range of scientific topics. See Beagon, *Roman Nature* (1992); French,

Ancient Natural History (1994); French and Greenaway (eds), *Science in the Early Roman Empire* (1986).

Studies Concentrating on Particular Areas of Inquiry

A word of caution: some of these areas of inquiry do not map exactly onto modern categories, even when the names (like 'physics' [*physika*]) seem the same.

Physics: Sambursky, *The Physical World of the Greeks* (1956); *The Physics of the Stoics* (1959); *The Physical World of Late Antiquity* (1962); Pedersen and Pihl, *Early Physics and Astronomy* (1974); Kahn, *Anaximander and the Origins of Greek Cosmology* (1960); Waterlow, *Nature, Change and Agency in Aristotle's Physics* (1982); Freudenthal, *Aristotle's Theory of Material Substance* (1995); Lang, *The Order of Nature in Aristotle's Physics* (1998); Furley, *Two Studies in the Greek Atomists* (1967), *The Greek Cosmologists* (1987); R. Sorabji, *Matter, Space and Motion* (1988).
Mechanics: Berryman, *The Mechanical Hypothesis in Ancient Greek Natural Philosophy* (2009); de Groot, *Aristotle's Empiricism* (2014).
Meteorology: Taub, *Ancient Meteorology* (2003); Wilson, *Structure and Method in Aristotle's* Meteorologica (2013).
Mathematics: Neugebauer *Exact Sciences in Antiquity* (1957, 2nd ed.); Cuomo, *Ancient Mathematics* (2001); Robson and Stedall (eds), *The Oxford Handbook to the History of Mathematics* (2009); see, particularly, chapters by Lloyd, Asper, Romano, Saito; Heath, *A History of Greek Mathematics* (1921); Knorr, *The Evolution of the Euclidean Elements* (1975); Netz, *The Shaping of Deduction in Greek Mathematics* (1999); Netz, *Ludic Proof* (2009); O'Meara, *Pythagoras Revived*; Dilke, *Mathematics and Measurement* (1987).
Astronomy: Neugebauer, *A History of Ancient Mathematical Astronomy* (1975, 3 vols); Evans, *The History and Practice of Ancient Astronomy* (1998); Pederson, *A Survey of the* Almagest (1974); Barton, *Ancient Astrology* (1994).
Harmonics: Barker, *The Science of Harmonics in Classical Greece* (2007); Creese, *The Monochord in Ancient Greek Harmonic Science* (2010).
Life Sciences: Pellegrin, *Aristotle's Classification of Animals*, trans. A. Preus (1986); Gotthelf and J. Lennox (eds), *Philosophical Issues in Aristotle's Biology* (1987); Leroi, *The Lagoon: How Aristotle Invented Science* (2014); Sedley, *Creationism and its Critics in Antiquity* (2007); Nutton, *Ancient Medicine* (2nd ed.); Totelin and Hardy, *Ancient Botany* (2015).
Geography: Dueck and Brodersen, *Geography in Classical Antiquity* (2012, in this series); Talbert and Brodersen (eds), *Space in the Roman World* (2004); Dilke, *Greek and Roman Maps* (1985).
Technology: Cuomo, *Technology and Culture in Greek and Roman Antiquity* (2007, in this series; see her Bibliographical Essay); Hannah, *Time in Antiquity* (2009).

Textual Studies of Greek and Roman Scientific, Mathematical, Medical and Technical Writings

The work of Kullmann and his school (including Kullmann, Althoff, and Asper [eds], *Gattungen wissenschaftlicher Literatur in der Antike*, 1998; Asper, *Griechische Wissenschaftstexte*, 2007; Föllinger, 'Fachliteratur 1. Gattungsbegriff und Gattungsgeschichte', 2011; Lengen, *Form und Funktion der aristotelischen Pragmatie*, 2002) has been particularly important, as are the contributions of a number of other scholars, including, for example, Fuhrmann, *Das systematische Lehrbuch* (1960); Fögen, *Wissen, Kommunikation und Selbstdarstellung* (2009). Conte, *Genres and Readers*, trans. Most (1994), van der Eijk, 'Towards a Rhetoric of Ancient Scientific Discourse' (1997) and Schenkeveld, 'Philosophical Prose' (1997) are useful starting points in English. The Introduction to *Structures and Strategies in Ancient Greek and Roman Technical Writing* (Doody, Föllinger and Taub, eds [2012]) discusses trends in the scholarship, particularly over the past twenty years. The following edited collections provide examples of recent scholarship: Horster and Reitz, *Antike Fachschriftsteller* (2003); Fögen, *Antike Fachtexte: Ancient Technical Texts* (2005); Taub and Doody, *Authorial Voices in Greco-Roman Technical Writing* (2009); Doody, Föllinger and Taub (2012, mentioned earlier); Asper (ed.), *Writing Science: Medical and Mathematical Authorship in Ancient Greece* (2013). Depew and Obbink (eds), *Matrices of Genre* (2000) is a valuable collection of essays by classicists interested in issues related to genre, not only in scientific and technical texts; see, for example, the chapter by Sluiter, 'The Dialectics of Genre' (2000). A number of scholars working in later periods are also particularly concerned with issues related to genre in medicine and philosophy, as well as scientific inquiries: see, for example, Pomata (2011) on 'epistemic' genres and selected chapters in Lavery and Groarke (eds) (2010), *Literary Form, Philosophical Content*.

On Literacy and 'Bookish' Culture

See Thomas, *Literacy and Orality in Ancient Greece* (1992, in this series) and her Bibliographical Essay; Harris, *Ancient Literacy* (1989); Yunis (ed.) *Written Texts and the Rise of Literate Culture in Ancient Greece* (2003), including chapters by Kahn and Dean-Jones (listed in the references). On cultures of reading and writing: Knox, 'Books and readers in the Greek world: from the beginnings to Alexandria' (1989); Easterling, 'Books and readers in the Greek world: the Hellenistic and Imperial Periods' (1989); Reynolds and Wilson, *Scribes and Scholars*, 4th edn (2013); Taub (2000).

On Specific Genres

Poetry: Toohey (1996); Volk (2002).
Dialogue: Nightingale (1995); Kahn (1996); Taub (2008), chap. 3.

Letters: Ceccarelli (2013); Morello and Morrison (eds) (2007); Rosenmeyer (2001); Trapp (2003).
Encyclopedia: Murphy (2004); Doody (2009).
Commentary: Tuominen (2009); Sorabji (ed.) (1990); Most (ed.) (1999).
Biography: Momigliano (1993); Hadas and Smith (1965); Meyer (1978); Warren (2007).
'Problem texts': Mayhew (ed.) (2015); De Leemans and Goyens (2006).
Recipes: Totelin (2009).
Doxography: While doxography may not be a genre, the following are worth consulting: Mansfeld and Runia (1997); van der Eijk (ed.) (1999).

References

Ancient Sources

Alexander of Aphrodisias. *Alexandri in Aristotelis Analyticorum priorum: librum I commentarium*. Ed. M. Wallies. Berlin. 1883.
Alexandri in Aristotelis Meteorologicorum libros commentaria. Ed. M. Hayduck. Berlin. 1899.
On Aristotle. Prior Analytics 1.32–46. Trans. I. Mueller. London. 2006.
Allen, H.W. (ed.) (1912) *Homeri Opera: Tomus V. Hymni, Cyclus, Fragmenta, Margites, Batrachomyomachia, Vitae*. Oxford.
[Apollodorus.] *Apollodorus' Library and Hyginus' Fabulae: Two Handbooks of Greek Mythology*. Trans. R.S. Smith and S.M. Trzaskoma. Indianapolis. 2007.
Apollodorus. *The Library*. Trans. J.G. Frazer. 2 vols. Loeb Classical Library. Cambridge, MA. 1921.
Apollonius of Perga. *Treatise on Conic Sections*. Ed. T.L. Heath. Cambridge. 1896.
Archimedes. 'Cattle Problem', in Thomas (trans.) 1939–1941. Vol. 2: 202–205.
Opera omnia. Ed. J.L. Heiberg. 2nd edn. 3 vols. Leipzig. 1910–1915.
The Works of Archimedes edited in modern notation with introductory chapters by T.L. Heath, with a supplement The Method of Archimedes, recently discovered by Heiberg. New York, n.d. (This Dover edition is a reissue of the 1897 edition and includes the Supplement of 1912, by arrangement with The Cambridge University Press.)
The Works of Archimedes: Translation and Commentary. Volume 1: The Two Books On the Sphere and the Cylinder. Trans. R. Netz. Cambridge. 2004.
[Aristotle.] *The Complete Works of Aristotle: The Revised Oxford Translation*. Ed. J. Barnes. 2 vols. Princeton. 1984.
Aristotle. *History of Animals*. Trans. D.M. Balme and A.L. Peck. 3 vols. Loeb Classical Library. Cambridge, MA. 2002.
History of Animals. Trans. D'A.W. Thompson. Oxford. (In *Complete Works*, ed. J. Barnes, 1984. Vol. 1: 774–993.)
Metaphysics. Trans. W.D. Ross. Oxford. (In *Complete Works*, ed. J. Barnes, 1984. Vol. 2: 1552–1728.)
Meteorologica. Trans. H.D.P. Lee. Loeb Classical Library. Cambridge, MA. 1952.

On the Heavens. Trans. W.K.C. Guthrie. Loeb Classical Library. Cambridge, MA. 1939.
On Sophistical Refutations. Trans. E.S. Forster. Loeb Classical Library. Cambridge, MA. 1955.
Physics. Trans. F.M. Cornford, and P.H. Wicksteed. Loeb Classical Library. Cambridge, MA. 1934.
Politics. Trans. H. Rackham. Loeb Classical Library. Cambridge, MA. 1944.
Topica. Trans. E.S. Forster. Loeb Classical Library. Cambridge, MA. 1960. (In volume with Posterior Analytics.)
Athenaeus. The Learned Banqueters. [Deipnosophists.] Trans. S.D. Olson. Loeb Classical Library. Cambridge, MA. 2006–2012.
Bacchylides. Greek Lyric: Bacchylides, Corinna and others. Vol. IV. Trans. D.A Campbell. Loeb Classical Library. Cambridge, MA. 1992.
Cicero. De Oratore III, De Fato, Paradoxa Stoicorum and De Partitione Oratoria. Trans. H. Rackham. Loeb Classical Library. Cambridge, MA. 1942.
  Letters to Friends. [Epistulae ad Familiares.], Vol. 1. Trans. D.R.S. Bailey. Loeb Classical Library. Cambridge, MA. 2001.
Cleomedes. Cleomedes' lectures on astronomy: a translation of The heavens. Trans. Alan C. Bowen and Robert B. Todd. Berkeley. 2004.
Columella. On Agriculture. Trans. H.B. Ash, E.S. Forster and E.H. Heffner. 3 vols. Loeb Classical Library. Cambridge, MA. 1955.
[Demetrius.] On Style. Trans. W.R. Roberts. Cambridge. 1902.
Diogenes Laertius. Lives of Eminent Philosophers. Trans. R.D. Hicks. 2 vols. Loeb Classical Library. Cambridge, MA. 1925.
Diophantus. In Diophanti Alexandri Opera Omnia. Ed. P. Tannery. 2 vols. Leipzig. 1895.
[Eratosthenes.] [Letter to King Ptolemy], in Eutocius, Commentary on Archimedes' Sphere and Cylinder, ed. J.L. Heiberg, in Archimedes Opera. 2nd edn. Leipzig. 1915: vol. 3, pp. 88–97.
  [Letter to King Ptolemy], Eutocius, Commentary on Archimedes' Sphere and Cylinder ii, trans. I. Thomas, in Selections Illustrating the History of Greek Mathematical Works. Loeb Classical Library. Cambridge, MA. 1939–1941. Vol. 1: 257–261 and 291–297. (The text from Eutocius is not completely reproduced here).
  'Letter to Ptolemy Euergetes', trans. I.E. Drabkin, in A Source Book in Greek Science, ed. M.R. Cohen and I.E. Drabkin. New York. 1948, pp. 62–66.
Euripides. Euripidis Tragoediae. Vol. 3. Perditarum tragoediarum fragmenta. Ed. A. Nauck. Leipzig. 1902.
Evelyn-White, H.G. (trans.) 1914. Hesiod. The Homeric Hymns and Homerica. Loeb Classical Library. Cambridge, MA.
Galen. Opera omnia. Ed. C.G. Kühn. Leipzig. 1821–1833.
  My Own Books, in Galen: Selected Works. Trans. P.N. Singer. Oxford. 2002, pp. 3–22.
  On My Own Opinions. Trans. and ed. V. Nutton. Berlin. 1999.
  The Greek Anthology. Trans. W.R. Paton. The Greek Anthology, Books XIII-XVI. Vol. 5. Loeb Classical Library. Cambridge, MA. 1918.

Gregory of Nazianzus. 'Letter to Nicobolus' (*Epistle* 51) in A.J. Malherbe, *Ancient Epistolary Theorists*. Atlanta. 1988.
[Herodotus] *Herodotus*. Trans. A.D. Godley. Loeb Classical Library. Cambridge, MA. 1922.
[Hesiod] *Hesiod. The Homeric Hymns and Homerica*. Trans. H.G. Evelyn-White. Loeb Classical Library. Cambridge, MA. 1914.
Hesiod *Theogony*. Trans. R. Lattimore. Ann Arbor. 1959.
Works and Days. Theogony. Trans. S. Lombardo. Indianapolis. 1993.
Hipparchus. *Ipparchou tōn Aratou kai Eudoxou phainomenōn exēgēseōs biblia tria = Hipparchi in Arati et Eudoxi phaenomena commentariorum libri tres*. Ed. C. Manitius. Leipzig. 1894.
[Hippocrates of Cos] *Hippocrates; Heracleitus. Nature of Man. Regimen in Health. Humours. Aphorisms. Regimen 1–3. Dreams. Heracleitus: On the Universe*. Trans. W.H.S. Jones. Loeb Classical Library. Cambridge, MA. 1931.
[Homer] *The Odyssey of Homer*. Trans. R.A. Lattimore. New York. 1975.
Homeric Hymns. Homeric Apocrypha. Lives of Homer. Trans. M.L. West. Loeb Classical Library. Cambridge, MA. 2003.
[Hyginus] *Apollodorus' Library and Hyginus' Fabulae: Two Handbooks of Greek Mythology*. Trans. R.S. Smith and S.M. Trzaskoma. Indianapolis. 2007.
Iamblichus. *Iamblichi De Communi Mathematica Scientia Liber*. Ed. N. Festa. Leipzig. 1891.
In Nicomachi Arithmeticam introductionem liber ad fidem codicis Florentini. Ed. H. Pistelli. Leipzig. 1894.
On the Pythagorean Way of Life. Texts and Translations 29, *Graeco-Roman Religion Series* 11. Trans. J. Dillon and J. Herschbell. Atlanta. 1991.
[Lucian] *Lucian. Volume I*. Trans. A.M. Harmon. Loeb Classical Library. Cambridge, MA. 1921.
Lucretius *On the nature of the universe*. Trans. R.E. Latham. Harmondsworth. 1994.
The nature of things. Trans. A.E. Stallings. London. 2007.
[Nicomachus] *Nicomachus' Arithmētikē Eisagōgē*. Ed. R. Hoche. Leipzig. 1866.
Olympiodorus. *Commentary on Plato's Gorgias*. Trans. R. Jackson, K. Lycos and H. Tarrant. Leiden. 1998.
Olympiodori in Aristotelis Meteora Commentaria. Ed. W. Stüve. Berlin. 1900.
Olympiodori philosophi In Platonis Gorgiam commentaria. Ed. W. Norvin. Leipzig. 1936.
Olympiodori in Platonis Gorgiam commentaria. Ed. L.G. Westerink. Berlin. 1970.
Philoponus, John. *Ioannis Philoponi in Aristotelis Analytica Priora Commentaria*. Ed. M. Wallies. Berlin. 1905.
Ioannis Philoponi in Aristotelis Meteorologicorum librum primum commentarium. Ed. M. Hayduck. Berlin. 1901.
On Aristotle On Coming-to-be and Perishing 2.5–11. Trans. I. Kupreeva. London. 2005.
On Aristotle Meteorology 1.1–3. Trans. I. Kupreeva. London. 2011.
On Aristotle Meteorology 1.4–9, 12. Trans. I. Kupreeva. London. 2012.

On Aristotle Physics 1.1–3. Trans. C. Osborne. London. 2006.
Plato. *Complete Works*. Ed. J.M. Cooper and D.S. Hutchinson. Indianapolis. 1997.
  Laws. In *Plato in Twelve Volumes*. Vols. 10 and 11. Trans. R.G. Bury. Loeb Classical Library. Cambridge, MA and London. 1926.
  Plato's Cosmology. [*Timaeus*] Trans. F.M. Cornford. London. 1935.
  Republic. Trans. C.J. Emlyn-Jones and W. Preddy. Loeb Classical Library. Cambridge, MA. 2013.
  Theaetetus; Sophist. Trans. H.N. Fowler. Loeb Classical Library. Cambridge, MA. 1921.
  Timaeus, Critias, Cleitophon, Menexenus and Epistles. Trans. R.G. Bury. Loeb Classical Library. Cambridge, MA. 1929.
Pliny the Elder. *Natural History*. 10 vols. Trans. H. Rackham. Loeb Classical Library. Cambridge, MA. 1938–1962.
Plutarch. *Table-talk* [*Quaestionum convivialium*]. In *Plutarch Moralia*, vol. 9. Trans. E.L. Minar, Jr. Loeb Classical Library. Cambridge, MA. 1961.
[Porphyry.] *Plotinus. Volume I: Porphyry On the Life of Plotinus and the Order of his Books*. Trans. A.H. Armstrong. Loeb Classical Library. Cambridge, MA. 1966.
Porphyry. *The Life of Pythagoras*. Trans. M. Smith, in M. Hadas and M. Smith, *Heroes and Gods: Spiritual Biographies in Antiquity*. London. 1965.
The Texts of Early Greek Philosophy: The Fragments and Selected Testimonies of the Major Presocratics. Trans. and ed. D. Graham. Cambridge. 2010.
Proclus. *Procli Diadochi in primum Euclidis Elementorum librum commentarii*. Ed. G. Friedlein. Leipzig. 1873.
Quintilian. *Institutio Oratoria*. Trans. H.E. Butler. Loeb Classical Library. Cambridge, MA. 1922.
Simplicius. *Simplicii in Aristotelis Categorias commentarium*. Ed. C. Kalbfleisch. Berlin. 1907.
  Simplicii in Aristotelis De Caelo commentaria. Ed. J.L. Heiberg. Berlin. 1894.
  Simplicii in Aristotelis Physicorum libros quattuor posteriores commentaria. Ed. H. Diels. Berlin. 1895.
Strabo. *Geography*. Trans. H.C. Hamilton and W. Falconer. London. 1903. http://data.perseus.org/citations/urn:cts:greekLit:tlg0099.tlg001.perseus-eng1:6.2.1. Accessed 20 July 2015.
  The Geography of Strabo. Trans. H.L. Jones. 8 vols. Loeb Classical Library. Cambridge, MA. 1932.
  The Geography of Strabo. Trans. D.W. Roller. Cambridge. 2014.
Suetonius. *C. Suetonii Tranquilli praeter Caesarum libros reliquiae. Pars prior. De grammaticis et rhetoribus*. Ed. G. Brugnoli. 2nd ed. Leipzig. 1963.
Thomas, I. (trans.) (1939–1941) *Selections Illustrating the History of Greek Mathematics*. 2 vols. Loeb Classical Library. Cambridge, MA.
Thucydides. *History of the Peloponesian War*. Trans. B. Jowett. Oxford. 1881. http://data.perseus.org/citations/urn:cts:greekLit:tlg0003.tlg001.perseus-eng2:6.2. Accessed 20 July 2015.
  Thucydides. Trans. C.F. Smith. Loeb Classical Library. Cambridge, MA. 1923.

Varro, Marcus Terentius. *On Agriculture*. Trans. W.D. Hooper. Loeb Classical Library. Cambridge, MA. 1935.

West, M.L. (trans.) (1993) *Greek Lyric Poetry: the poems and fragments of the Greek iambic, elegiac, and melic poets (excluding Pindar and Bacchylides) down to 450 BC*. Oxford.

(ed. and trans.) (2003) *Homeric Hymns; Homeric Apocrypha; Lives of Homer*. Loeb Classical Library. Cambridge, MA.

Modern Sources

Albiani, M.G. (2006) 'Metrodorus', in H. Cancik and H. Schneider (eds), *Brill's New Pauly: Encyclopaedia of the Ancient World*. Leiden, Vol. VIII: 839 (9).

Aly, W. (1929) '*Sphragis* (1)', in A. Pauly, G. Wissowa, and W. Kroll (eds), *Real-Encyclopädie der classischen Altertumswissenschaft*. Stuttgart, Bd. III.2, cols 1757–1758.

Amthor, A. (1880) 'Das Problema bovinum des Archimedes'. *Zeitschrift für Mathematik und Physik* 25: 153–171.

Andersen, O. (1987) 'Mündlichkeit und Schriftlichkeit im frühen Griechentum'. *Antike und Abendland* 33: 29–44.

Armstrong, A.H. (1966) *Plotinus. Volume I: Porphyry On the Life of Plotinus and the Order of his Books*. Loeb Classical Library. Cambridge, MA.

Asmis, E. (1984) *Epicurus' Scientific Method*. Ithaca, NY.

Asper, M. (2007) *Griechische Wissenschaftstexte. Formen, Funktionen, Differenzierungsgeschichten*. Stuttgart.

(2009) 'The Two Cultures of Mathematics in Ancient Greece', in E. Robson and J. Stedall (eds) *The Oxford Handbook of the History of Mathematics*. Oxford, pp. 107–32.

(ed.) (2013) *Writing Science: Medical and Mathematical Authorship in Ancient Greece*. Berlin.

Aujac, G. (2001) *Eratosthène de Cyrène, le pionnier de la géographie: sa mesure de la circonférence terrestre*. Paris.

Baltussen, H. (2007) 'Playing the Pythagorean: Ion's *Triagmos*', in Jennings and Katsaros (eds) (2007), pp. 295–318.

(2008) *Philosophy and Exegesis in Simplicius*. London.

Bar-Kochva, B. (2010) *The Image of the Jews in Greek Literature: The Hellenistic Period*. Berkeley.

Barker, A. (2007) *The Science of Harmonics in Classical Greece*. Cambridge.

(2012). 'Aristoxenus', in Hornblower et al. (eds) (2012).

Barnes, J. (1987) *Early Greek Philosophy*. Harmondsworth.

(1992) 'Diogenes Laertius IX 61–116: The Philosophy of Pyrrhonism', in *Aufstieg und Niedergang der Römischen Welt*. Berlin, Bd. II.36.6: 4241–4301.

(2000). *Aristotle: A Very Short Introduction*. Oxford.

Barton, T. (1994) *Ancient Astrology*. London.

Beagon, M. (1992) *Roman Nature: The Thought of Pliny the Elder*. Oxford.

Benson, G.C. (2014) 'Archimedes the Poet: Generic Innovation and Mathematical Fantasy in the Cattle Problem'. *Arethusa* 47.2: 169–196.
Berggren, J.L. and A. Jones (2000) *Ptolemy's Geography: an Annotated Translation of the Theoretical Chapters*. Princeton and Oxford.
Berryman, S. (2009) *The Mechanical Hypothesis in Ancient Greek Natural Philosophy*. Cambridge.
Bickerman, E.J. (1980) *Chronology of the Ancient World*. Revised edn. London.
Bing, P. (1998) 'Between Literature and the Monuments', in M.A. Harder, R.F. Regtuit, and G.C. Wakker (eds), *Genre in Hellenistic Poetry*. Groningen, pp. 21–43.
Bing, P. and J.S. Bruss (eds) (2007) *Brill's Companion to Hellenistic Epigram*. Leiden.
Blair, A. (1999) 'The Problemata as a Natural Philosophical Genre', in A. Grafton and N. Siraisi (eds), *Natural Particulars: Nature and the Disciplines in Renaissance Europe*. Cambridge, MA, pp. 171–204.
Blümer, W. (2001) *Interpretation archaischer Dichtung: die mythologischen Partien der Erga Hesiods. Vol. I: Die Voraussetzungen: Autoren, Texte und homerische Fragen*. Münster.
Bodéüs, R. (1993) *The Political Dimensions of Aristotle's Ethics*. Trans. J.E. Garrett. Albany, NY.
Bowie, E. (2007) 'From Archaic Elegy to Hellenistic Sympotic Epigram?', in Bing and Bruss (eds) (2007), pp. 95–112.
Boyer, C.B. (1968) *A History of Mathematics*. New York.
(1987). *The Rainbow: From Myth to Mathematics*. Basingstoke.
Boys-Stones, G.R., B. Graziosi, and P. Vasunia (eds) (2009) *The Oxford Handbook of Hellenic Studies*. Oxford.
Broadie, S. (2012) *Nature and Divinity in Plato's Timaeus*. Cambridge.
Bruss, J.S. (2010).'Epigram', in J.J. Clauss and M. Cuypers (eds), *A Companion to Hellenistic Literature*. Chichester, pp. 117–135.
Burkert, W. (1972). *Lore and Science in Ancient Pythagoreanism*. Trans. E.L. Minar, Jr. Cambridge, MA. [Original publication: *Weisheit und Wissenschaft: Studien zu Pythagoras, Philolaos und Platon*, Nürnberg, 1962.]
Burnyeat, M. (2005). 'Archytas and Optics'. *Science in Context* 18.1: 35–53.
Burnyeat, M. and M. Frede (2015). *The Pseudo-Platonic Seventh Letter: A Seminar*. Ed. D. Scott. Oxford.
Burridge, R.A. (1992/1995) *What Are the Gospels? A Comparison with Graeco-Roman Biography*. Cambridge.
Cameron, A. (1967) 'The End of the Ancient Universities'. *Cahiers d'Histoire Mondiale* 10: 653–673.
(1993) *The Greek Anthology from Meleager to Planudes*. Oxford.
(1995) *Callimachus and His Critics*. Princeton.
Cancik, H. and H. Schneider (eds) (2002–2010) *Brill's New Pauly: Encyclopaedia of the Ancient World*. Leiden.
Carey, C. (2007) 'Epideictic Oratory', in I. Worthington (ed.), *A Companion to Greek Rhetoric*. Malden, MA and Oxford, pp. 236–252.

Carey, S. (2003) *Pliny's Catalogue of Culture: Art and Empire in the Natural History*. Oxford.
Ceccarelli, P. (2013) *Ancient Greek Letter Writing: A Cultural History (600 BC-150 BC)*. Oxford.
Chase, M. (2003) *On Aristotle's 'Categories 1–4'*: Simplicius. Ithaca, NY.
Chibnall, M. (1975) 'Pliny's Natural History and the Middle Ages', in T.A. Dorey (ed.), *Empire and Aftermath*. London, pp. 57–78.
Christianidis, J. (1994) 'On the History of Indeterminate Problems of the First Degree in Greek Mathematics', in K. Gavroglu, J. Christianidis, and E. Nicolaidis (eds), *Trends in the Historiography of Science*. London, pp. 237–247.
Clackson, J. (2015) *Language and Society in the Greek and Roman Worlds*. Cambridge.
Clagett, M. (1999) *Ancient Egyptian Science: A Source Book. Volume 3. Ancient Egyptian Mathematics*. Philadelphia.
  (1957) *Greek Science in Antiquity*. London.
Clark, G. (1989) *Iamblichus: On the Pythagorean Life* (*Translated Texts for Historians* 8). Liverpool.
Clarke, D.M. (2011) 'Hypotheses', in D.M. Clarke and C. Wilson (eds), *The Oxford Handbook of Philosophy in Early Modern Europe*. Oxford, pp. 249–271.
Clay, D. (1983) *Lucretius and Epicurus*. Ithaca, NY.
  (1990) *The Philosophical Inscriptions of Diogenes Oenoanda: New Discoveries 1969–1983*. Berlin.
  (2009) 'The Athenian Garden', in J. Warren (ed.), *The Cambridge Companion to Epicureanism*. Cambridge, pp. 9–28.
Clements, A. (2014) *Aristophanes' Thesmophoriazusae: Philosophizing Theatre and the Politics of Perception in Late Fifth-Century Athens*. Cambridge.
Codoñer, C. (1991) 'De l'antiquité au moyen âge: Isidore de Seville', in A. Becq (ed.), *L'encyclopédisme: Actes du colloque de Caen 12–16 janvier 1987*. Paris, pp. 19–35.
Cohen, M.R. and I.E. Drabkin (1948) *A Source Book in Greek Science*. Cambridge, MA.
Combès, J. (1986) 'Introduction', in J. Combès and L.G. Westerink (eds), *Damascius, Traité des premiers principes*. Paris, Vol. I, pp. ix–lxxii.
Conte, G.B. (1994) *Genres and Readers: Lucretius, Love Elegy, Pliny's Encyclopedia*. Trans. G.W. Most. Baltimore and London.
Conte, G.B. and Most, G.W. (2012) 'Genre', in Hornblower et al. (eds) (2012).
Copenhaver, B.P. (1978) 'The Historiography of Discovery in the Renaissance: The Sources and Composition of Polydore Vergil's *De inventoribus rerum*, I-III'. *Journal of the Warburg and Courtauld Institutes* 41: 192–214.
Cooper, J.M. and D.S. Hutchinson (eds) (1997) *Plato: Complete Works*. Indianapolis.
Craik, E. (2006) 'Horizontal Transmission in the Hippocratic Tradition'. *Mnemosyne* 59.3: 334–347.

Creese, D. (2010) *The Monochord in Ancient Greek Harmonic Science.* Cambridge.
Cribiore, R. (1996) *Writing, Teachers, and Students in Greco-Roman Egypt.* Atlanta.
  (2001) *Gymnastics of the Mind: Greek Education in Hellenistic and Roman Egypt.* Princeton.
Cunningham, A. (1988) 'Getting the Game Right: Some Plain Words on the Identity and Invention of Science'. *Studies in History and Philosophy of Science* 19.3: 365–389.
Cuomo, S. (2001) *Ancient Mathematics.* London.
  (2007) *Technology and Culture in Greek and Roman Antiquity.* Cambridge.
Darwin, E. (1791) *The Botanic Garden: A Poem in Two Parts.* London.
De Groot, J. (2014) *Aristotle's Empiricism: Experience and Mechanics in the Fourth Century BC.* Las Vegas.
De Leemans, P. and M. Goyens (2006) *Aristotle's Problemata in Different Times and Tongues.* Leuven.
Dean-Jones, L. (2003) 'Literacy and the Charlatan in Ancient Greek Medicine', in Yunis (ed.) (2003), pp. 97–121.
Diels, H. (1879). *Doxographi Graeci.* Berlin.
Delatte, A. (1922) *La Vie de Pythagore de Diogène Laërce. Académie Royale de Belgique, Classe des Lettres et des Sciences Morales et Politiques, Mémoires. Deuxième Série* 17: 3–270.
Depew, M. and D. Obbink (eds) (2000) *Matrices of Genre: Authors, Canons, and Society.* Cambridge, MA.
Dickey, E. (2007) *Ancient Greek Scholarship.* Oxford.
Dilke, O.A.W. (1985) *Greek and Roman Maps.* London.
  (1987) *Mathematics and Measurement.* London.
Dillon, J. and J. Herschbell (1991) *Iamblichus: On the Pythagorean Way of Life.* (*Texts and Translations* 29, Graeco-Roman Religion Series 11). Atlanta.
Dirlmeier, F. (1962) *Merkwürdige Zitate in der Eudemischen Ethik des Aristoteles.* Heidelberg.
Doody, A. (2009) 'Pliny's *Natural History*: Enkuklios Paideia and the Ancient Encyclopedia'. *Journal of the History of Ideas* 70.1: 1–21.
  (2010) *Pliny's Encyclopaedia: The Reception of the Natural History.* Cambridge.
Doody, A., S. Föllinger, and L. Taub (eds) (2012) *Structures and Strategies in Ancient Greek and Roman Technical Writing.* (*Studies in History and Philosophy of Science* 43, 2: 233–325; Introduction 233–236).
Dover, K.J. (1997) *The Evolution of Greek Prose Style.* Oxford.
Dueck, D. and K. Brodersen (2012) *Geography in Classical Antiquity.* Cambridge.
Duff, D. (ed.) (2000) *Modern Genre Theory.* London.
Dzielska, M. (1986) *Apollonius of Tyana in Legend and History.* Trans. P. Pieńkowski. Rome.
  (1995) *Hypatia of Alexandria.* Cambridge, MA.
Easterling, P.E. (1989) 'Books and Readers in the Greek World: the Hellenistic and Imperial Periods', in P.E. Easterling and B.M.W. Knox (eds), *The Cambridge History of Classical Literature.* Cambridge, Vol. I part 4, pp. 169–197.

Easterling, P.E. and B.M.W. Knox (eds) (1989) *The Cambridge History of Classical Literature*. Cambridge, Vol. I part 4.
Eastwood, B. and G. Graßhoff (2004) *Planetary Diagrams for Roman Astronomy in Medieval Europe CA. 800–1500*. Philadelphia.
[Einstein, Albert] (1949) *Albert Einstein: Philosopher-Scientist*. Ed. P.A. Schilpp. Evanston, IL.
Empson, W. (1930) *Seven Types of Ambiguity*. London.
Enk, P.J. (1970) 'Encyclopaedic Learning', in N.G.L. Hammond and H.H. Scullard (eds), *Oxford Classical Dictionary*. 2nd edn. Oxford, p. 383.
Evans, J. (1998) *The History and Practice of Ancient Astronomy*. Oxford.
  (1999) 'The Material Culture of Greek Astronomy'. *Journal for the History of Astronomy* 30: 237–307.
  (2005) '*Gnômonikê Technê*: The Dialer's Art and Its Meaning in the Ancient World', in W. Orchiston (ed.), *The New Astronomy: Opening the Electromagnetic Window and Expanding Our View of the Planet Earth*. New York, pp. 273–292.
Evans. J.A.S. (1978) 'What Happened to Croesus?' *The Classical Journal* 74.1: 34–40.
Évrard, E. (1957) *L'École d'Olympiodore et la composition du Commentaire à la Physique de Jean Philopon*. Dissertation. Liège.
Falcon, A. (2005) *Aristotle and the Sciences of Nature*. Cambridge.
Farrington, B. (1939) *Science and Politics in the Ancient World*. London.
  (1944) *Greek Science*. Harmondsworth. (Originally published as *Greek Science: Its Meaning for Us*. Part I (1944); reprinted with Part II 1953; revised one-volume edn 1961).
Fischer, K. (2013) 'Der Begriff der "(wissenschaftlichen) Abhandlung" in der griechischen Antike – eine Untersuchung des Wortes πραγματεία'. *Antike Naturwissenschaft und ihre Rezeption* 23: 93–114.
Flinterman, J.-J. (1995) *Power, Paideia & Pythagoreanism: Greek Identity, Conceptions of the Relationship between Philosophers and Monarchs and Political Ideas in Philostratus' Life of Apollonius*. Amsterdam.
Fögen, T. (2005) *Antike Fachtexte: Ancient Technical Texts*. Berlin.
  (2009) *Wissen, Kommunikation und Selbstdarstellung: zur Struktur und Charakteristik römischer Fachtexte der frühen Kaiserzeit*. Munich.
Föllinger, S. (2011) 'Fachliteratur I: Gattungsbegriff und Gattungsgeschichte', in B. Zimmermann (ed.), *Die Literatur der archaischen und klassischen Zeit*. (*Handbuch der griechischen Literatur der Antike* 7). Munich, Bd. 1, pp. 289–292.
  (2012) 'Aristotle's Biological Works as Scientific Literature', in Doody et al. (eds) (2012), pp. 237–244.
Folkerts, M. (1978) *Die älteste mathematische Aufgabensammlung in lateinischer Sprache: die Alkuin zugeschriebenen Propositiones ad acuendos iuvenes*. Critical edn. Vienna.
  (2000) 'Mesolabion', in *Der Neue Pauly: Enzyklopädie der Antike*. Stuttgart and Weimar, Bd. VIII: 17.

Fontaine, R. (1995) *Otot ha-Shamayim: Samuel Ibn Tibbon's Hebrew Version of Aristotle's Meteorology*. Leiden.
Forster, E.S. (1945) 'Riddles and Problems from the Greek Anthology'. *Greece and Rome* 14: 42–47.
Fortenbaugh, W.W., P.M. Huby and A.A. Long (eds) (1985) *Theophrastus of Eresus: On His Life and Work*. New Brunswick.
Fowler, D.H. (1980) *Archimedes Cattle Problem and the Pocket Calculating Machine*. Coventry.
  (1999) *The Mathematics of Plato's Academy: A New Reconstruction*. 2nd edn. Oxford.
Fraser, P.M. (1972) *Ptolemaic Alexandria*. 3 vols. Oxford.
Fraser, P. (2012) 'Eratosthenes', in Hornblower et al. (eds) (2012).
Frede, M. (1974) 'Stoic vs. Aristotelian Syllogistic'. *Archiv für Geschichte der Philosophie* 56: 1–32.
Freeland, C.A. (1990) 'Scientific Explanation and Empirical Data in Aristotle's *Meteorology*'. *Oxford Studies in Ancient Philosophy* 8: 67–102.
French, R. (1994) *Ancient Natural History: Histories of Nature*. London.
French, R. and F. Greenaway (eds) (1986) *Science in the Early Roman Empire: Pliny the Elder, His Sources and Influence*. London.
Freudenthal, G. (1995) *Aristotle's Theory of Material Substance*. Oxford.
Fuhrmann, M. (1960) *Das systematische Lehrbuch: ein Beitrag zur Geschichte der Wissenschaften in der Antike*. Göttingen.
Furley, D. (1967) *Two Studies in the Greek Atomists*. Princeton.
  (1987) *The Greek Cosmologists: The Formation of the Atomic Theory and Its Earliest Critics*. Cambridge.
Galilei, G. (1632) *Dialogo sopra i due massimi sistemi del mondo*. Florence.
  (1638) *Discorsi e dimostrazioni matematiche intorno a due nuove scienze*. Leiden.
Gardiner, F. (2003) 'Diogenes Laertius: The Man Behind the Text'. Undergraduate thesis. Cambridge.
Gärtner, H.A. (2008) 'Riddles', in Cancik and Schneider (eds) (2008) Vol. XII: 587–591. (http://referenceworks.brillonline.com/entries/brill-s-new-pauly/riddles-e1018330; accessed 14 May 2014).
Gee. E. (2013) *Aratus and the Astronomical Tradition*. Oxford.
Genette, G. (1997) *Paratexts: Thresholds of Interpretation*. Trans. J.E. Lewin. Cambridge. (Originally published as *Seuils*. Paris 1987).
Geus, K. (2002) *Eratosthenes von Kyrene: Studien zur hellenistischen Kultur- und Wissenschaftsgeschichte*. Munich.
  (2004) 'Measuring the Earth and the *Oikoumene*: Zones, Meridians, *Sphragides* and Some Other Geographical Terms used by Eratosthenes of Cyrene', in R. Talbert and K. Brodersen (eds), *Space in the Roman World: Its Perception and Presentation*. Münster, pp. 11–26.
Gibson, R.K. and C.S. Kraus (eds) (2002) *The Classical Commentary: Histories, Practices, Theory*. Leiden.
Gill, C., T. Whitmarsh and J. Wilkins (eds) (2009) *Galen and the World of Knowledge*. Cambridge.

Gill, M.L. and P. Pellegrin (eds) (2006) *A Companion to Ancient Philosophy*. Malden, MA and Oxford.
Goldhill, S. (2002) *The Invention of Prose*. Oxford.
Gorman, P. (1979) *Pythagoras: A Life*. London.
Gotthelf, A. and J. Lennox (eds) (1987) *Philosophical Issues in Aristotle's Biology*. Cambridge.
Gow, A.S.F. and D.L. Page (1965) *The Greek Anthology: The Hellenistic Epigrams*. 2 vols. Cambridge.
  (1968) *The Greek Anthology: The Garden of Philip*. 2 vols. Cambridge.
Goyens, M., P. De Leemans and A. Smets (eds) (2008) *Science Translated: Latin and Vernacular Translations of Scientific Treatises in Medieval Europe*. Leuven.
Graf, F. (2012) 'Pythagoras (1), Pythagoreanism: Religious Aspects of Pythagoreanism', in Hornblower et al. (eds) (2012).
Grafton, A. and M. Williams (2006) *Christianity and the Transformation of the Book*. Cambridge, MA.
Graham, D.W. (2010) *The Texts of Early Greek Philosophy: The Complete Fragments and Selected Testimonies of the Major Presocratics*. Cambridge.
Graßhoff, G. (1990) *The History of Ptolemy's Star Catalogue*. New York.
Greaves, M. (2002) *Philosophical Status of Diagrams*. Stanford, CA.
Griffin, M. (trans.) (2015) 'Introduction' to *Olympiodorus: Life of Plato* and *On Plato First Alcibiades 1–9*. London.
Gutzwiller, K.J. (1998) *Poetic Garlands: Hellenistic Epigrams in Context*. Berkeley.
Hadas, M. and M. Smith (1965) *Heroes and Gods: Spiritual Biographies in Antiquity*. London.
Hadot, P. (1987) *Philosophy as a Way of Life: Spiritual Exercises from Socrates to Foucault*. Ed. A.I. Davidson; trans. M. Chase. Oxford.
Hallyn, F. (1990) *The Poetic Structure of the World: Copernicus and Kepler*. Trans. D.M. Leslie. New York.
Hamblin, C.L. (1976) 'An Improved Pons Asinorum?' *Journal of the History of Philosophy* 14.2: 131–136.
Hamilton, N.T., N.M. Swerdlow, and G.J. Toomer (1987) 'The Canobic Inscription: Ptolemy's Earliest Work', in J.L. Berggren and B.R. Goldstein (eds), *From Ancient Omens to Statistical Mechanics*. Copenhagen, pp. 55–73.
Hannah, R. (2009) *Time in Antiquity*. London.
Harder, M.A., R.F. Regtuit, and G.C. Wakker (eds) (2002) *Hellenistic Epigrams*. Groningen.
Harley, J.B. and D. Woodward (eds) (1987) *The History of Cartography. Volume I: Cartography in Prehistoric, Ancient and Medieval Europe and the Mediterranean*. Chicago.
Harris, W.V. (1989) *Ancient Literacy*. Cambridge, MA.
Harvey, P. (ed.) (1937/1974) *The Oxford Companion to Classical Literature*. Oxford.
Hayduck, M. (1899) *Commentaria in Aristotelem Graeca. Volume III.1–2: Alexandri in Aristotelis Meteorologicorum Libros Commentaria*. Berlin.
Healy, J.F. (1991) *Pliny the Elder: Natural History. A Selection*. Harmondsworth.

Heath, T.L. (1912) *The Works of Archimedes edited in modern notation with introductory chapters by T.L. Heath, with a supplement The Method of Archimedes, recently discovered by Heiberg*. New York. (This Dover edition is an unabridged reissue of the Heath edition of 1897 and includes the Supplement of 1912, by special arrangement with The Cambridge University Press.).
 (1921/1981) *A History of Greek Mathematics*. 2 vols. Oxford.
 (1956) *The Thirteen Books of Euclid's Elements*. New York.
 (1964) *Diophantus of Alexandria: A Study in the History of Greek Algebra*. New York.
 (1981) *Aristarchus of Samos: The Ancient Copernicus*. New York. (Originally published Oxford, 1913.)
Heiberg, J.L. (1910–1915) *Archimedes: Opera Omnia*. 2nd edn. 3 vols. Leipzig.
Hermann, C.F. (1878) *Platonis Dialogi VI*. Leipzig.
Hine, H. (2009) 'Subjectivity and Objectivity in Latin Scientific and Technical Literature', in Taub and Doody (eds) (2009), pp. 13–30.
Hope, R. (1930) *The Book of Diogenes Laertius, Its Spirit and Its Method*. New York.
Hopkins, B.C. (2011) *The Origin of the Logic of Symbolic Mathematics*. Bloomington, IN.
Hornblower, S., A. Spawforth, and E. Eidinow (eds) (2012) *The Oxford Classical Dictionary*. 4th edn. Oxford.
Horster, M. and C. Reitz (eds) (2003) *Antike Fachschriftsteller: literarischer Diskurs und sozialer Kontext*. Stuttgart.
Høyrup, J. (1997) 'Hero, Ps.-Hero, and Near Eastern Practical Geometry: an Investigation of *Metrica, Geometrica* and other Treatises', in K. Döring, B. Herzhoff, and G. Wöhrle (eds), *Antike Naturwissenschaft und ihre Rezeption* 7. Trier, pp. 67–93.
 (2001) 'On a Collection of Geometric Riddles and Their Role in the Shaping of Four to Six <<Algebras>>'. *Science in Context* 14: 85–131.
 (2006) 'Bronze Age Formal Science? With Additional Remarks on the Historiography of Distant Mathematics', in B. Löwe, V. Peckhaus, and T. Räsch (eds), *Foundations of the Formal Sciences IV: The History of the Concept of the Formal Sciences*. London, pp. 81–102.
Huffman, C.A. (2005) *Archytas of Tarentum: Pythagorean, Philosopher and Mathematician King*. Cambridge.
Hume, D. (1739–1740) *A Treatise of Human Nature*. London.
 (1779) *Dialogues concerning Natural Religion*. London.
Humphrey, J.W., J.P. Oleson, and A.N. Sherwood (2002) *Greek and Roman Technology: A Sourcebook*. London.
Immerwahr, H.R. (1960) '*Ergon*: History as a Monument in Herodotus and Thucydides'. *American Journal of Philology* 81.3: 261–290.
Irby-Massie, G.L. and P.T. Keyser (2002) *Greek Science of the Hellenistic Era: A Sourcebook*. London.
Jackson, H. (1920) 'Aristotle's Lecture Room and Lectures'. *Journal of Philology* 35: 191–200.
Jaeger, M. (2008) *Archimedes and the Roman Imagination*. Ann Arbor.

Jardine, N. and A. Segonds (1999) 'Kepler as Reader and Translator of Aristotle', in C. Blackwell and S. Kusukawa (eds), *Philosophy in the Sixteenth and Seventeenth Centuries: Conversations with Aristotle*. Aldershot, pp. 206–33.
Jennings, V. and A. Katsaros (eds) (2007) *The World of Ion of Chios*. Leiden.
Johansen, T.K. (2004) *Plato's Natural Philosophy: A Study of the Timaeus-Critias*. Cambridge.
Jones, A. and L. Taub (eds) (forthcoming) *The Cambridge History of Science*, Vol. I. Cambridge.
Jouanna, J. (1999) *Hippocrates*. Trans. M.B. DeBevoise. Baltimore and London. (Originally published as *Hippocrate*. Paris: 1992.)
Judson, L. (1991) *Aristotle's Physics: a Collection of Essays*. Oxford.
Kahn, C.H. (1960) *Anaximander and the Origins of Greek Cosmology*. New York
  (1996) *Plato and the Socratic Dialogue: The Philosophical Use of a Literary Form*. Cambridge.
  (2001) *Pythagoras and the Pythagoreans: A Brief History*. Indianapolis and Cambridge.
  (2003) 'Writing Philosophy: Prose and Poetry from Thales to Plato', in Yunis (ed.), pp. 139–161.
Kearns, E. (1989) *The Heroes of Attica*. (*Bulletin of the Institute of Classical Studies*, Supplement 57). London.
  (2012) 'Hero-cult', in Hornblower et al. (eds) (2012).
Kepler, J. (1870) *Opera Omnia*, Vol. VIII. Ed. Ch. Frisch. Frankfurt.
  (1945) *Briefe 1590–1599*, in M. Caspar (ed.), *Gesammelte Werke*, Bd. XIII, Munich.
  (1951) *Johannes Kepler: Life and Letters*. Trans. C. Baumgardt. New York.
Keyser, P.T. and G.L. Irby-Massie (eds) (2008) *The Encyclopedia of Ancient Natural Scientists: The Greek Tradition and Its Many Heirs*. London.
Kirk, G.S., J.E. Raven, and M. Schofield (1983) *The Presocratic Philosophers*. 2nd edn. Cambridge.
Klein, J. (1968) *Greek Mathematical Thought and the Origin of Algebra*. Cambridge, MA.
Kneale, M. and W. Kneale (1962) *The Development of Logic*. Oxford.
Knorr, W.R. (1975) *The Evolution of the Euclidean Elements: A Study of the Theory of Incommensurable Magnitudes and Its Significance for Early Greek Geometry*. Dordrecht.
  (1986) *The Ancient Tradition of Geometric Problems*. Boston.
  (1989) *Textual Studies in Ancient and Medieval Geometry*. Boston.
Knox, B.M.W. (1989) 'Books and Readers in the Greek World: From the Beginnings to Alexandria', in P.E. Easterling and B.M.W. Knox (eds), *The Cambridge History of Classical Literature*. Cambridge, Vol. I, part 4, pp. 154–168.
König, J. and T. Whitmarsh (eds) (2007) *Ordering Knowledge in the Roman Empire*. Cambridge.
Konstan, D. (2007) 'Cartesian Solitude and the Nature of Reading'. (Unpublished paper presented at the conference 'Philosophy and Literature: Reading across the Disciplines', Wesleyan University, CT, May 9–10, 2007.)

Kranz, W. (1961) 'Sphragis: Ichform und Namensiegel als Eingangs- und Schlussmotiv Antiker Dichtung'. *Rheinisches Museum für Philologie* 104: 3–46, 97–124.
Kühnert, F. (1961) *Allgemeinbildung und Fachbildung in der Antike*. Berlin.
Kullmann, W., J. Althoff, and M. Asper (eds) (1998) *Gattungen wissenschaftlicher Literatur in der Antike*. Tübingen.
Kupreeva, I. (2005) *Philoponus: On Aristotle On Coming-to-be and Perishing 2.5–11*. London.
  (2011) *Philoponus: On Aristotle Meteorology 1.1–3*. London.
  (2012) *Philoponus: On Aristotle Meteorology 1.4–9, 12*. London.
Kurke, L. (2010) *Aesopic Conversations: Popular Tradition, Cultural Dialogue, and the Invention of Greek Prose*. Princeton.
Kwapisz, J. (2013) 'Were there Hellenistic Riddle Books?', in J. Kwapisz, D. Petrain, and M. Szymanski (eds), *The Muse at Play: Riddles and Wordplay in Greek and Latin Poetry*. Berlin, pp. 148–167.
Laks, A. (2008) 'Speculating about Diogenes of Apollonia', in P. Curd and D.W. Graham (eds), *The Oxford Handbook of Presocratic Philosophy*. Oxford, pp. 353–364.
Lamberton, R. (2001) *Plutarch*. New Haven.
Lamberton, R. and J.J. Keaney (1992) *Homer's Ancient Readers: The Hermeneutics of Greek Epic's Earliest Exegetes*. Princeton.
Lang, H.S. (1998) *The Order of Nature in Aristotle's Physics: Place and the Elements*. Cambridge.
Lavery, J. and L. Groarke (eds) (2010) *Literary Form, Philosophical Content: Historical Studies of Philosophical Genres*. Madison, WI.
Law, V. and I. Sluiter (1998) *Dionysius Thrax and the Technē Grammatikē*. Münster.
Lee, H.D.P. (trans.) (1952) *Aristotle: Meteorologica*. Loeb Classical Library. Cambridge, MA.
Lengen, R. (2002) *Form und Funktion der aristotelischen Pragmatie: die Kommunikation mit dem Rezipienten*. Stuttgart.
Lennox, J.G. (2001) *Aristotle's Philosophy of Biology: Studies in the Origins of Life Sciences*. Cambridge.
Leroi, A.M. (2014) *The Lagoon: How Aristotle Invented Science*. London.
Lessing, G.E. (1773) *Zur Geschichte und Litteratur: Aus den Schätzen der Herzoglichen Bibliothek zu Wolfenbüttel. Zweyter Beytrag. Bd. XIII: Zur griechischen Anthologie*. Braunschweig. (http://diglib.hab.de/drucke/lo-4637-1b-2s/start.htm/; accessed 26 July 2015).
Lettinck, P. (1999) *Aristotle's Meteorology and its Reception in the Arab World. With an Edition and Translation of Ibn Suwār's Treatise on Meteorological Phenomena and Ibn Bājja's Commentary on the Meteorology*. Leiden.
Leventhal, M. (2013) 'Between Poetry and Mathematics in the Hellenistic Age: The Epic-Grammatic Case of Archimedes' Cattle Problem'. MPhil. Thesis. Cambridge.
  (2015) 'Counting on Epic: Mathematical Poetry and Homeric Epic in Archimedes' *Cattle Problem*'. *Ramus* 44.1–2: 200–21.

Lévy, I. (1927) *La Légende de Pythagore de Grèce en Palestine*. Paris.
Liddell, H.G., R. Scott, and H.S. Jones (1940) *A Greek-English Lexicon*. 9th edn. (With a Supplement 1968). Oxford.
Livingstone, N. and G. Nisbet (2010) *Epigram* (*Greece and Rome New Surveys in the Classics* 38). Cambridge.
Lloyd, G.E.R. (1966) *Polarity and Analogy in Greek Thought*. Cambridge. (Repr. Bristol 1987).
  (1970) *Early Greek Science: Thales to Aristotle*. London.
  (1973) *Greek Science after Aristotle*. London.
  (1978) 'The Empirical Basis of the Physiology of the *Parva Naturalia*', in G.E.R Lloyd and G.E.L. Owen (eds), *Aristotle on Mind and the Senses*. Cambridge, pp. 215–239.
  (1979) *Magic, Reason and Experience: Studies in the Origins and Development of Greek Science*. Cambridge.
  (1983) *Science, Folklore and Ideology*. Cambridge.
  (1991) 'Introduction' to 'Observational error in later Greek science', in Lloyd (1991), pp. 299–302.
  (1991) *Methods and Problems in Greek Science: Selected Papers*. Cambridge.
  (1998) *Aristotelian Explorations*. Cambridge.
Long, H.S. (1925) 'Introduction' to R.D. Hicks (trans.), *Diogenes Laertius: Lives of Eminent Philosophers*. Loeb Classical Library. 2 vols. Cambridge, MA, Vol. I, pp. xv–xxvi.
  (2012) 'Diogenes (6) Laertius', in Hornblower et al. (eds) (2012). Revised by R.W. Sharples.
Lord, A. (1960) *The Singer of Tales*. New York.
Malherbe, A.J. (1988) *Ancient Epistolary Theorists*. Atlanta.
Mansfeld, J. (1999) 'Sources', in K. Algra, J. Barnes, J. Mansfeld, and M. Schofield (eds), *The Cambridge History of Hellenistic Philosophy*. Cambridge, pp. 3–30.
Mansfeld, J. and D. Runia (1997) *Aëtiana: the Method and Intellectual Context of a Doxographer*. Leiden.
Marrou, H.I. (1956) *A History of Education in Antiquity*. Trans. G. Lamb. New York.
Mattern, S.P. (2013) *The Prince of Medicine: Galen in the Roman Empire*. Oxford.
Mayhew, R. (ed.) (2015) *The Aristotelian Problemata Physica: Philosophical and Scientific Investigations*. Leiden.
Mayor, A. (2010) *The Poison King: The Life and Legend of Mithradates, Rome's Deadliest Enemy*. Princeton.
McLuhan, M. (1964) *Understanding Media: The Extensions of Man*. New York.
Meissner, B. (1999) *Die technologische Fachliteratur der Antike: Struktur, Überlieferung und Wirkung technischen Wissens in der Antike (ca. 400 v. Chr.-ca.500 n. Chr.)*. Berlin.
Mejer, J. (1978) *Diogenes Laertius and his Hellenistic Background*. (*Hermes Einzelschriften* 40). Wiesbaden.
Midonick, H.O. (1965) *The Treasury of Mathematics: a Collection of Source Material in Mathematics Edited and Presented with Introductory Biographical and Historical Sketches*. New York.

Momigliano, A. (1993) *The Development of Greek Biography*. Expanded edn. Cambridge, MA.
Morello, R. and A.D. Morrison (eds) (2007) *Ancient Letters: Classical and Late Antique Epistolography*. Oxford.
Morgan, T. (1998) *Literate Education in the Hellenistic and Roman Worlds*. Cambridge.
Morrow, G.R. (1960) *Plato's Cretan City: A Historical Interpretation of the Laws*. Princeton.
  (1962) *Plato's Epistles: a Translation with Critical Essays and Notes*. Indianapolis.
  (trans.) (1970/1992) *Proclus: A Commentary on the First Book of Euclid's Elements*. Princeton.
Mosley, A. (2007) *Bearing the Heavens: Tycho Brahe and the Astronomical Community of the Late Sixteenth Century*. Cambridge.
Moss, J.D. (1993) *Novelties in the Heavens: Rhetoric and Science in the Copernican Controversy*. Chicago.
Most, G.W. (ed.) (1999) *Commentaries – Kommentare*. Göttingen.
  (2000) "Generating Genres: The Idea of the Tragic", in Depew and Obbink (eds) (2000), pp. 15–35.
Mourelatos, Alexander P.D. (1970/2008) *The Route of Parmenides: A Study of Word, Image and Argument in the Fragments*. New Haven; rev. edn, Las Vegas.
Mueller, I. (1974) 'Greek Mathematics and Greek Logic', in J. Corcoran (ed), *Ancient Logic and Its Modern Interpretations*. Dordrecht, pp. 35–70.
  (1992) 'Foreword to the 1992 edition', in Morrow (1970/1992), pp. ix–xxxi.
Mugler, C. (1958) *Dictionnaire historique de la terminologie géométrique des Grecs*. Paris.
Murphy, T. (2004) *Pliny the Elder's Natural History: The Empire in the Encyclopaedia*. Oxford.
Naiden, F.S. (2013) *Smoke Signals for the Gods: Ancient Greek Sacrifice from the Archaic through Roman Periods*. Oxford.
Nauert, C. (1979) 'Humanists, Scientists and Pliny: Changing Approaches to a Classical Author'. *American Historical Review* 84: 72–85.
  (1980) 'Caius Plinius Secundus', in F.E. Crantz and P.O. Kristeller (eds), *Catalogus Translationum et Commentariorum: Medieval and Renaissance Latin Translations and Commentaries*. Washington, DC, pp. 297–422.
Netz, R. (1999a) *The Shaping of Deduction in Greek Mathematics*. Cambridge.
  (1999b) 'Proclus' Division of the Mathematical Proposition into Parts: How and Why Was It Formulated?'. *Classical Quarterly* 49.1: 282–303.
  (2002) 'Greek Mathematicians: a Group Picture', in Tuplin and Rihll (eds) (2002), pp. 196–216.
  (2004) *The Works of Archimedes. Volume 1: The Two Books On the Sphere and the Cylinder*. Cambridge.
  (2009a) *Ludic Proof: Greek Mathematics and the Alexandrian Aesthetic*. Cambridge.
  (2009b) 'The Exact Sciences', in Boys-Stones et al. (eds) (2009), pp. 579–596.
Netz, R. and W. Noel (2007) *The Archimedes Codex: How a Medieval Prayer Book Is Revealing the True Genius of Antiquity's Greatest Scientist*. London.

Neugebauer, O. (1957) *The Exact Sciences in Antiquity*. 2nd edn. Providence, RI.
  (1975) *A History of Ancient Mathematical Astronomy*. 3 vols. Berlin.
Newton, I. (1687) *Philosophiae naturalis principia mathematica*. London.
Nightingale, A.W. (1995) *Genres in Dialogue: Plato and the Construct of Philosophy*. Cambridge.
Nisbet, G. (2003) *Greek Epigram in the Roman Empire: Martial's Forgotten Rivals*. Oxford.
Nutton, V. (2009) 'Galen's Authorial Voice: A Preliminary Enquiry', in Taub and Doody (eds) (2009), pp. 53–62.
  (2013) *Ancient Medicine*. 2nd edn. London.
Ohlert, C. (1912) *Rätsel und Rätselspiele der alten Griechen*. 2nd edn. Berlin. (Reprint: Miami n.d.).
Olding, G. (2007) 'Shot from the Canon: Sources, Selections, Survival', in Jennings and Katsaros (eds) (2007), pp. 45–63.
O'Meara, D.J. (1989) *Pythagoras Revived: Mathematics and Philosophy in Late Antiquity*. Oxford.
  (2012) 'Iamblichus (2)', in Hornblower et al. (eds) (2012).
Osborne, C. (1998) 'Was Verse the Default Form for Presocratic Philosophy?', in C. Atherton (ed.), *Form and Content in Didactic Poetry*. Bari, pp. 23–35.
  (2006) *Philoponus: On Aristotle Physics 1.1–3*. London.
Page, D.L. (1981) *Further Greek Epigrams*. Cambridge.
Palmer, J.A. (2008/2012) 'Parmenides', in E.N. Zalta (ed.), *The Stanford Encylopedia of Ancient Philosophy*. Stanford, CA. (http://plato.stanford.edu/entries/parmenides/; accessed 2 May 2015).
  (2009) *Parmenides and Presocratic Philosophy*. Oxford.
Paltridge, B. (1997) *Genre, Frames and Writing in Research Settings*. Amsterdam and Philadelphia.
Parry, M. (1971) *The Making of Homeric Verse: The Collected Papers of Milman Parry*. Ed. A. Parry. Oxford.
Paton, W.R. (1918) *The Greek Anthology.* 5 vols. Loeb Classical Library. London.
Patterson, C. (2013) 'Education in Plato's Laws', in J.E. Grubbs, T. Parkin, and R. Bell (eds), *The Oxford Handbook to Childhood and Eduction in the Classical World*. Oxford, pp. 365–380.
Pedersen, O. (1974) *A Survey of the Almagest*. Odense.
  (1986) 'Some Astronomical Topics in Pliny', in R.K. French and F. Greenaway (eds) *Science in the Early Roman Empire: Pliny the Elder, His Sources and Influences*. London, pp. 162–196.
Pedersen, O. and M. Pihl (1974) *Early Physics and Astronomy*. London.
Pellegrin, P. (1986) *Aristotle's Classification of Animals: Biology and the Conceptual Unity of the Aristotelian Corpus*. Trans. A. Preus. Berkeley. (Originally published as *La classification des animaux chez Aristote: statut de la biologie et unité de l'aristotélisme*. Paris 1982.)
Pelling, C. (2007) 'Ion's *Epidemiai* and Plutarch's Ion', in Jennings and Katsaros (eds) (2007), pp. 75–109.
  (2012) 'Biography, Greek', in Hornblower et al. (eds) (2012).

Pfeiffer, R. (1968) *History of Classical Scholarship from the Beginnings to the End of the Hellenistic Age.* Oxford.

Pomata, G. (2011) 'Observation Rising: Birth of an Epistemic Genre, ca. 1500–1650', in L. Daston and E. Lunbeck (eds), *Histories of Scientific Observation.* Chicago, pp. 45–80.

Praechter, K. (1927) 'Simplikios' in A. Pauly, G. Wissowa, and W. Kroll (eds), *Real-Encyclopädie der classischen Altertumswissenschaft.* Stuttgart, Bd. IIIA.1, cols 204–213.

(1990) 'Review of the *Commentaria in Aristotelem Graeca*', in Sorabji (ed.) (1990), pp. 31–54. (Originally published as 'Die griechischen Aristoteleskommentare,' *Byzantinische Zeitschrift* 18 (1909): 516–38.)

Pryce, F.N., D.E. Strong, and M. Vickers (2012) 'Seals', in Hornblower et al. (eds) (2012).

Purcell, N. (2012) 'Pliny (1) the Elder', in Hornblower et al. (eds) (2012).

Reynolds, L.D. (1983) *Texts and Transmission: a Survey of the Latin Classics.* Oxford.

Reynolds, L.D. and N.G. Wilson (2013) *Scribes and Scholars: a Guide to the Transmission of Greek and Latin Literature.* 4th edn. Oxford.

Rheinberger, H.-J. (2003) 'Discourse of Circumstances: A Note on the Author in Science', in M. Biagioli and P. Galison (eds), *Scientific Authorship: Credit and Intellectual Property in Science.* London, pp. 309–324.

Rihll, T.E. (1999) *Greek Science.* Oxford.

Roberts, C.H. and T.C. Skeat (1983) *The Birth of the Codex.* Oxford.

Roberts, D.H. (2012) '*Sphragis*', in Hornblower et al. (eds) (2012).

Robson, E. and J. Stedall (eds) (2009) *The Oxford Handbook of the History of Mathematics.* Oxford.

Rokem, F. (1996) 'One Voice and Many Legs: Oedipus and the Riddles of the Sphinx', in G. Hasan-Roken (ed.), *Untying the Knot: On Riddles and Other Enigmatic Modes.* Oxford, pp. 255–270.

Roller, D.W. (2010) *Eratosthenes' Geography.* Princeton.

Rosenmeyer, P. (2001) *Ancient Epistolary Fictions: The Letter in Greek Literature.* Cambridge.

(2006) *Ancient Greek Literary Letters: Selections in Translation.* London.

Rosso, M. (2008) 'User-based Identification of Web Genres'. *Journal of the American Society for Information Science and Technology* 59.7: 1053–1072.

Rowe, D.E. and S. Unguru (1981) 'Does the Quadratic Equation Have Greek Roots? A Study of "Geometric Algebra", "Applications of Areas" and Related Problems'. *Libertas Mathematica* 1: 1–49.

Runia, D.T. (1999) 'What is Doxography?', in P.J. van der Eijk (ed.), *Ancient Histories of Medicine: Essays in Medical Doxography and Historiography in Classical Antiquity.* (*Studies in Ancient Medicine* 20). Leiden, pp. 33–55.

Russell, D.A.F.M. (2012) 'Demetrius (17)', in Hornblower et al. (eds) (2012).

Sambursky, S. (1956) *The Physical World of the Greeks.* London.

(1959) *The Physics of the Stoics.* London.

(1962) *The Physical World of Late Antiquity.* Princeton.

Schenkeveld, D.M. (1997) 'Philosophical Prose', in S.E. Porter (ed.), *Handbook of Classical Rhetoric in the Hellenistic Period 330 B.C.-A.D. 400*. Leiden, pp. 195–264.
Schiesaro, A. (2012) 'Didactic Poetry', in Hornblower et al. (eds) (2012).
Schiesaro, A., P. Mitsis, and J.S. Clay (eds) (1993) *Mega Nepios: il destinatario nell' epos didascalico*. (Materiali e discussioni per l'analisi dei testi classici 31.) Pisa.
Schultz, W. (1914) 'Rätsel', in A. Pauly, G. Wissowa, and W. Kroll (eds), *Real-Encyclopädie der classischen Altertumswissenschaft*. Stuttgart, Bd. I, cols 62–125.
Scolnicov, S. (2004) 'Plato', in P.S. Fass (ed.), *Encyclopedia of Children and Childhood in History and Society*. New York, pp. 681–682. (www.faqs.org/childhood/Pa-Re/Plato-427-348-B-C-E.html/).
Sedley, D.N. (1998) *Lucretius and the Transformation of Greek Wisdom*. Cambridge.
  (2007) *Creationism and Its Critics in Antiquity*. Berkeley.
Sesiano, J. (2004) 'Introduction', in J. Christianidis (ed.), *Classics in the History of Greek Mathematics*. London, pp. 257–63.
Sharples, R.W. (1987) 'Alexander of Aphrodisias: Scholasticism and Innovation', in *Aufstieg und Niedergang der römischen Welt*. Berlin, Bd. II.36.2, pp. 1176–1243.
  (1990) 'The School of Alexander?', in Sorabji (ed.) (1990), pp. 83–111.
  (1994) *Alexander of Aphrodisias: Quaestiones 2.16–3.15*. London.
  (2012) 'Alexander (14) (*RE* 94) of Aphrodisias', in Hornblower et al. (eds) (2012).
Sider, D. (2005) *The Fragments of Anaxagoras*. Sankt Augustin.
Sidoli, N. (2015) 'Mathematics Education', in W.M. Bloomer (ed.), *A Companion to Ancient Education*. Hoboken, NJ, pp. 387–400.
Sluiter, I. (2000) 'The Dialectics of Genre: Some Aspects of Secondary Literature and Genre in Antiquity', in M. Depew and D. Obbink (eds), *Matrices of Genre: Authors, Canons and Society*. Cambridge, MA, pp. 183–203.
Smith, A. (2012) 'Porphyry', in Hornblower et al. (eds) (2012).
Smith, M.F. (1993) *The Epicurean Inscription: Diogenes of Oinoanda*. Naples.
Smith, M. (1996) *The Philosophical Inscriptions of Diogenes of Oinoanda*. Vienna.
  (2003) *Supplement to Diogenes of Oinoanda The Epicurean Inscription*. Naples.
Smith, R. (1978) 'The Mathematical Origins of Aristotle's Syllogistic'. *Archive for History of Exact Sciences* 19: 201–209.
  (1989) *Aristotle: Prior Analytics. Translated, with Introduction, Notes and Commentary*. Indianapolis.
Smith, W. (ed.) (1867) *A Dictionary of Greek and Roman Biography and Mythology*. Vol. II. Boston. (http://name.umdl.umich.edu/ACL3129.0002.001/; accessed 23 March 2015)
Snyder, G. (2000) *Teachers and Texts in the Ancient World: Philosophers, Jews and Christians*. London.
Söderqvist, T. (2007) '"No Genre of History Fell Under More Odium Than That of Biography": The Delicate Relations Between Scientific Biography and

Historiography of Science', in T. Söderqvist (ed.), *The History and Poetics of Scientific Biography*. Aldershot, pp. 241–262.

Solmsen, F. (1960) *Aristotle's System of the Physical World: A Comparison with his Predecessors*. Ithaca, NY.

Sorabji R. (1988) *Matter, Space and Motion*. London.

(1990) 'The ancient commentators on Aristotle', in Sorabji (ed.) (1990), pp. 1–30.

(ed.) (1990) *Aristotle Transformed: the Ancient Commentators and Their Influence*. London.

Soutar, G. (1939) *Nature in Greek Poetry: Studies Partly Comparative*. Oxford.

Swales, J.M. (1990) *Genre Analysis: English in Academic and Research Settings*. Cambridge.

Symons, G.J. and J.G. Wood (eds) (1894) *Theophrastus of Eresus on Winds and Weather Signs*. London.

Taavitsainen, I. (2001) 'Changing Conventions of Writing: The Dynamics of Genres, Text Types, and Text Traditions'. *European Journal of English Studies* 5.2: 139–150.

Talbert, C.H. (1977) *What is a Gospel? The Genre of the Canonical Gospels*. Philadelphia.

Talbert, R. and K. Brodersen (eds) (2004) *Space in the Roman World: Its Perception and Presentation*. Münster.

Tatarkiewicz, W. (1963) 'Classification of the Arts in Antiquity'. *Journal of the History of Ideas* 24.2: 231–240.

Taub, L. (1993) *Ptolemy's Universe: The Natural Philosophical and Ethical Foundations of His Astronomy*. Chicago.

(2000) 'Ancient Science', in A. Hunter (ed.), *Thornton and Tully's Scientific Books, Libraries and Collectors*. 4th edn. Aldershot, pp. 26–71.

(2002) 'Instruments of Alexandrian Astronomy: The Uses of the Equinoctial Rings', in C.J. Tuplin and T.E. Rihll (eds.), *Science and Mathematics in Ancient Greek Culture*. Oxford, pp. 133–149.

(2003) *Ancient Meteorology*. London.

(2007) 'Presenting a "Life" as a Guide to Living: Ancient Accounts of the Life of Pythagoras', in Thomas Söderquist (ed.), *The History and Poetics of Scientific Biography*. Aldershot, pp. 17–36.

(2008a) *Aetna and the Moon: Explaining Nature in Ancient Greece and Rome*. Corvallis, OR.

(2008b) '"Eratosthenes Sends Greetings to King Ptolemy": Reading the Contents of a "Mathematical" Letter'. *Acta Historica Leopoldina* 54: 285–302.

(2010) 'Translating the Phainomena Across Genre, Language and Culture', in A. Imhausen und T. Pommerening (eds), *Writings of Early Scholars in the Ancient Near East, Egypt and Greece: Zur Übersetzbarkeit von Wissenschaftssprachen des Altertums*. Berlin, pp. 119–137.

(2013) 'On the Variety of "Genres" of Greek Mathematical Writing: Thinking about Mathematical Texts and Modes of Mathematical Discourse', in Asper (ed.) (2013), pp. 333–65.

(2017) 'Archiving Science in Greco-Roman Antiquity', in L. Daston (ed.), *Science in the Archives: Pasts, Presents, and Future*. Chicago, pp. 113–35.
Taub, L. and A. Doody (eds) (2009) *Authorial Voices in Greco-Roman Technical Writing*. Trier.
Taylor, T. (1807) *The Treatises of Aristotle, On the Heavens, On Generation and Corruption, and On Meteors*. London.
Thesleff, H. (1949) 'Some Remarks on Literary *Sphragis* in Greek Poetry'. *Eranos* 47: 116–128.
Thillet, P. (trans.) (2008) *Aristote: Météorologiques*. Paris.
Thomas, R. (1992) *Literacy and Orality in Ancient Greece*. Cambridge.
(2003) 'Prose Performance Texts: *Epideixis* and Written Publication in the Late Fifth and Early Fourth Centuries', in Yunis (ed.) (2003), pp. 162–188.
Tiede, D.L. (1972) *The Charismatic Figure as Miracle Worker*. Missoula, MT.
Toohey, P. (1996) *Epic Lessons: an Introduction to Ancient Didactic Poetry*. London.
Toomer, G.J. (2012) 'Aratus (1)', in Hornblower et al. (eds) (2012).
Totelin, L. (2004) 'Mithradates' Antidote: a Pharmacological Ghost'. *Early Science and Medicine* 9: 1–19.
(2009) *Hippocratic Recipes: Oral and Written Transmission of Pharmacological Knowledge in Fifth- and Fourth-Century Greece*. Leiden.
Totelin, L. and G. Hardy (2015) *Ancient Botany*. London.
Trapp, M.B. (2003) *Greek and Latin Letters: an Anthology*. Cambridge.
(2012) 'Letters, Greek', in Hornblower et al. (eds) (2012).
Tueller, M.A. and R.T. Macfarlane (2009) 'Hipparchus and the Poets: a Turning Point in Scientific Literature', in M.A. Harder, R.F. Regtuit and G.C. Wakker (eds), *Nature and Science in Hellenistic Poetry*. Leuven, pp. 227–253.
Tuominen, M. (2009) *The Ancient Commentators on Plato and Aristotle*. Stocksfield, Northumberland.
Tuplin, C.J. and T. Rihll (eds) (2002) *Science and Mathematics in Ancient Greek Culture*. Oxford.
Unguru, S. (1975) 'On the Need to Rewrite the History of Greek Mathematics'. *Archive for the History of Exact Sciences* 15: 67–114.
Van der Eijk, P.J. (1997) 'Towards a Rhetoric of Ancient Scientific Discourse: Some Formal Characteristics of Greek Medical and Philosophical Texts (Hippocratic Corpus, Aristotle)', in E.J. Bakker (ed.), *Grammar as Interpretation: Greek Literature in its Linguistic Contexts*. Leiden, pp. 77–129.
(ed.) (1999) *Ancient Histories of Medicine: Essays in Medical Doxography and Historiography in Classical Antiquity*. Leiden.
Van der Waerden, B.L. (1954) *Science Awakening*. Trans. A. Dresden. Groningen.
Van Winden, J.C.M. (1959) *Calcidius On Matter, His Doctrines and Sources: A Chapter in the History of Platonism*. Leiden.
Vardi, I. (1998) 'Archimedes' *Cattle Problem*'. *The American Mathematical Monthly* 105: 305–319.
Verrycken, K. (1990) 'The Development of Philoponus' Thought and Its Chronology', in Sorabji (ed.) (1990), pp. 233–274.

(2010) 'John Philoponus', in L.P. Gerson (ed.), *The Cambridge History of Philosophy in Late Antiquity*, Volume II. Cambridge, pp. 733–755.
Vlastos, G. (1975) *Plato's Universe*. Oxford.
Volk, K. (2002) *The Poetics of Latin Didactic: Lucretius, Vergil, Ovid, Manilius*. Oxford.
Von Staden, H. (2002) '"A Woman does not become Ambidextrous": Galen and the Culture of the Scientific Commentary', in Gibson and Kraus (eds) (2002), pp. 109–139.
Wallace-Hadrill, A. (1990) 'Pliny the Elder and Man's Unnatural History'. *Greece & Rome*: 37.1: 80–96.
Warren, J. (2007) 'Diogenes Laërtius, Biographer of Philosophy', in König and Whitmarsh (eds) (2007), pp. 133–149.
Waszink, J.H. (1964) *Studien zum Timaioskommentar des Calcidius*. Leiden.
Waszink, J.H. and P.J. Jensen (eds) (1962) *Plato Latinus*. Volume IV: *Timaeus a Calcidio Translatus Commentarioque Instructus*. Leiden.
Waterlow, S. (1982) *Nature, Change and Agency in Aristotle's Physics: a Philosophical Study*. Oxford.
Watts, E.J. (2008) *City and School in Late Antique Athens and Alexandria*. Berkeley.
Werlich, E. (1976) *A Text Grammar of English*. Heidelberg.
West, M.L. (1966) *Hesiod Theogony: Edited with Prolegomena and Commentary*. Oxford.
(1978) *Hesiod Works and Days: Edited with Prolegomena and Commentary*. Oxford.
(1982) *Greek Metre*. Oxford.
(1985) 'Ion of Chios'. *Bulletin of the Institute of Classical Studies* 32: 71–78.
(1987) *Introduction to Greek Metre*. Oxford.
(2003) *Homeric Hymns. Homeric Apocrypha. Lives of Homer*. Loeb Classical Library. Cambridge, MA. 2003.
(2012) 'Riddles', in Hornblower et al. (eds) (2012).
Westerink, L.G. (1990) 'The Alexandrian Commentators and the Introductions to their Commentaries', in Sorabji (ed.) (1990), pp. 325–348.
Whewell, W. (1840) *The Philosophy of the Inductive Sciences Founded upon Their History*. 2 vols. London and Cambridge.
Whitehead, A.N. (1925) *Science and the Modern World: Lowell Lectures, 1925*. New York.
Wilamowitz-Moellendorff, U. von (1894/1971) 'Ein Weihgeschenk des Eratosthenes'. *Nachrichten der K. Gesellschaft der Wissenschaften zu Göttingen, Phil.-hist. Klasse* (1894): 15–35. (Reprinted in *Kleine Schriften*. Berlin 1971, Bd. II (*Hellenistische, spätgriechische und lateinische Poesie*), pp. 48–70.)
Wilcox, A. (2012) *The Gift of Correspondence in Classical Rome: Friendship in Cicero's Ad Familiares and Seneca's Moral Epistles*. Madison, WI.
Wildberg, C. (1988) *John Philoponus' Criticism of Aristotle's Theory of Aether*. Berlin.
(2008) 'John Philoponus', in E.N. Zalta (ed.), *The Stanford Encyclopedia of Philosophy*. Fall 2008 edn. Stanford, CA. (http://plato.stanford.edu/archives/fall2008/entries/philoponus/).

(2010) 'Prolegomena to the Study of Philoponus' *Contra Aristotelem*', in R. Sorabji (ed.), *Philoponus and the Rejection of Aristotelian Science*. 2nd edn. London, pp. 239–250.
Williams, H.C., R.A. German, and C.R. Zarnke (1965) 'Solution of the Cattle Problem of Archimedes'. *Mathematics of Computation* 19: 671–674.
Wilson, M. (2013) *Structure and Method in Aristotle's Meteorologica: A More Disorderly Nature*. Cambridge.
Worthington, I. (ed.) (2007) *A Companion to Greek Rhetoric*. Malden, MA and Oxford.
Wright, M.R. (1998) 'Philosopher Poets: Parmenides and Empedocles', in C. Atherton (ed.), *Form and Content in Didactic Poetry*. Bari, pp. 1–22.
Wurm, J.F. (1830) Review of J.G. Hermann, *De Archimedis Problemate Bovino*. *Jahrbücher für Philologie und Paedagogik* 14: 194–202.
Yatsuhashi, A. (2010) 'In the Bird Cage of the Muses: Archiving, Erudition, and Empire in Ptolemaic Egypt'. PhD dissertation. Duke University.
Yunis, H. (ed.) (2003) *Written Texts and the Rise of Literate Culture in Ancient Greece*. Cambridge.
Zeller, E. (1880–1892) *Die Philosophie der Griechen in ihrer geschichtlichen Entwicklung*, Bd. I-III. Leipzig.
Zhmud, L. (1998) 'Plato as Architect of Science?'. *Phronesis* 43.3: 211–244.
  (2001) 'Revising Doxography: Hermann Diels and His Critics.' *Philologus* 145: 219–43.
  (2006) *The Origin of the History of Science in Classical Antiquity*. Berlin.

Index

Aetna (anonymous) 24
Against the Christians (Porphyry) 126
agriculture
  animal behaviour and plants, importance
    of 83, 85
  astronomy, and 80–83, 85
  calendars/almanacs 81–82, 83, 84
  farmer as exemplar of ideal Roman 85
  Hesiod
    astronomy, and 80, 81–82
    value of agricultural life 83
  Natural History (Pliny), in 79, 80–83, 85
  On Agriculture (Columella) 23
  prediction as part of agricultural skill 79
  value of agricultural life 83
Alexander of Aphrodisias
  commentary 87
    Aristotle's *Meteorology*, on 92, 93, 96, 100,
      103–4, 107, 109–10
    Aristotle's *Physics*, on 93
    Aristotle's works, on 86–87, 92, 93
    written for publication 89
  teacher, as 93
Alexander Polyhistor
  Pythagorean memoirs/notebooks 120
Ammaeus 68
Ammonius 89, 93
Analytica (Aristotle) 93
Anaximander of Miletus
  first Greek prose treatise written
    by 10–11, 13
  Peri physeōs (*On nature*) 10–11, 13, 15
Anaximenes 121
Apollodorus
  The Library 25
Apollonius of Perga 61
  Conics 61, 62
  specialist mathematical texts subject of
    commentaries 89
Apollonius of Tyana

Life of Apollonius (Philostratus) 112–13
  marvels credited to 112–13
Apollonius Rhodius 56–57
apple (*mēlon*)
  calculational problems, use in 45–46
  counting of 48
  'sheep', and 45–46
Aratus of Soli
  astronomy 1, 3
  Phaenomena 3, 18, 29
    Hipparchus' commentary on 88
    Eudoxus' earlier prose work as source 29
  poetry 1
    astronomical 3
    didactic 24, 31
    prose antecedents of 29
Archimedes 19, 33, 34–35
  Cattle Problem see Cattle Problem (*problema
    bovinum*)
  Dositheus, and 63
  Eratosthenes, and
    competitor to 59
    corresponding with 35, 49, 51, 63, 64
    relationship between 38–39
  letters
    characteristic features of 64
    prefaces to technical mathematical
      text, as 63
    scientific and mathematical topics, on 52
  mathematical poetry 49
  Method concerning Mechanical Theorems 35
  Sand-Reckoner 34, 61
    parallel to *Letter to King Ptolemy*, as 62
  On the Sphere and the Cylinder 55, 61
    Eutocius' commentary on 88
  war machines, work on 59
Archytas of Tarentum 58
  Aristoxenus' biography of 112
Aristotle 15
  Analytica

181

Index

Aristotle (*cont.*)
  Philoponus' commentary on 93
  biography reflecting intellectual and ethical
    concerns, generating interest in 112
  Categories 87
    Philoponus' commentary on 93
  commentaries on works of 20, 89–90
    Alexander 86–87, 92, 93, 96, 100–10
    Olympiodorus 86, 92, 93, 98–99, 100–10
    Philoponus 88, 89–92, 93, 96–98, 100–10
    Simplicius 87, 88, 89
    target texts, Aristotle's works as 87
  de Anima
    Philoponus' commentary on 93
  diagrams in *Meteorology*, use of 92, 94, 100–10
    lettered diagrams 100–2, 104 fig. 4.1, 106–7
    mathematical approach 103–8
    Milky Way 102
    optics, study of 101–2
    shooting stars 106–7
    visual aids for lectures, diagrams as 100–1
    wind direction 102–3
    winds, treatment of 100, 102–3
  History of Animals 100–1
  knowledge, nature and types of 8–9
  lectures, work beginning as 132
  mathematics as a branch of theoretical
    knowledge 8
  Metaphysics
    Alexander's commentary on 93
    Philoponus' commentary on 93
  Meteorology 20
    Alexander's commentary on 92, 93,
      96, 100–10
    commentators adding own
      ideas/criticisms 93, 109
    diagrams, use of 92, 94, 100–10
    etymology of words, discussing 98–99
    exhalations 80
    explanatory tactics used by Aristotle
      92, 94–110
    Homer 28
    Olympiodorus' commentary on 92, 93,
      98–99, 100–10
    'open' character of genre of commentary
      95, 99–100, 109–10
    Philoponus' commentary on 92, 93,
      96–98, 100–10
    variety of sources/authorities, references to
      92, 94–100, 109
  Natural History, as authority in 80
  On Generation and Corruption
    Philoponus' commentary on 93
  On the heavens 88
  Physics 88
    Alexander's commentary on 93
    '*Lecture Course on Nature*', as 132
    Philoponus' commentary on 90–92
    Porphyry's commentary on 115–16
  Posterior Analytics 9–10
  pragmateia 16, 17
  Pythagoreans 115
  style of Greek scientific explanation 9–10
  Thales' monopoly in olive presses 111
Aristoxenus of Tarentum
  biographies of four philosophers
    Archytas 112
    Plato 112
    Pythagoras 112, 115–16, 119
    Socrates 112
Artemon of Cassandreia
  Aristotle's correspondence, publishing 51
Asper, Markus 49
Astronomica (Manilius) 24, 30
astronomy 1, 3
  agriculture, and 80–83, 85
  Astronomica (Manilius) 24, 30
  authorial choice of formats 1
  commentary on 88
  Natural History (Pliny), in
    astronomical knowledge, insufficiency
      of 83, 85
    astrometeorological calendars, questioning
      usefulness of 83
    difficulties of 81
    farming, and 80–81, 82–83, 85
    trade and navigation, and 80–81
    weather prediction 84
  On the heavens (Aristotle) 88
Athenaeus of Naucratis 25–26

Banquet of the Learned (*Deipnosophists*) 25–26
Beagon, Mary 74–75
Bing, Peter 31
biography 21, 111–29
  aims of 112, 114
  Archytas, *bioi* of 112
  eulogies or *encomia*, and 111–12
  heroes
    aretalogical accounts of heroic teachers as
      moral exemplars 113–14, 127
    divinity of heroic teacher, promoting 128
    'hero', meaning of 113, 114
    heroes, types of honours offered to 113
    heroic character of individuals, *bioi*
      emphasising 112–14
  intellectual and ethical concerns, reflecting 114
    Aristotle generating interest in 112
    bioi as standard form 112
  Life of Apollonius (Philostratus) 112–13

Index

Life of Epicurus (Diogenes Laertius) 19, 50–51
  nature of 111–12
  not a clearly demarcated genre for ancient Greeks/Romans 111–12
  On the Life of Plotinus (Porphyry) 86, 90
  Plato, *bioi* of 112
  Pythagoras, *bioi* of 112, 114–16
    Diogenes Laertius *bios* 21, 114, 116–21
    Iamblichus' *On the Pythagorean Life* 21, 111, 114, 123–29
    Porphyry's *Life of Pythagoras* 21, 114, 121–23
    purposes of 128–29
  Socrates, *bioi* of 112
  usual content of 117
Bowie, Ewen 48–49
Boyer, Carl B. 41
Burkert, Walter 115
Burridge, Richard A. 117

Calcidius
  commentary on Plato's *Timaeus* 88
Carey, Christopher 5
Carey, Sorcha 73, 75–77
Categories see under Aristotle
Cato 81
  encyclopaedic work/subjects divided into *artes* 74
Cattle Problem (*problema bovinum*) 22, 24, 32, 35–39, 46, 63
  authorship 34–35
  Eratosthenes and Archimedes, relationship between 38–39, 59
  myths, allusions to 46–47
  poetry, uses of 49
  very large number as solution to 37
Celsus
  encyclopaedic work/subjects divided into *artes* 74
Charmides (Plato) 44–46
children
  education through play 40–41, 44
  letter-writing 63
Christianity 108, 109–10, 126
  influence in shaping *bioi* of Pythagoras 121
  letters as important form of communication 51, 52, 70
  Porphyry's critique of 126
Chrysippus 117–18
Cicero
  comments on style 63–64
  letters 70
  Letters to Atticus 70
Clackson, James 130
Clark, Gillian 124–25

Claudius Ptolemy *see* Ptolemy (Claudius Ptolemy)
Clearchus of Soli
  On Riddles (*Peri griphōn*) 26
Cleomedes
  lectures 1
  work beginning as 132
  The Heavens 132
Cohen, Morris R. 9, 32
Columella
  agricultural life, value of 83
  On Agriculture 23
commentary 1, 2, 11, 20, 86–110
  Aristotle's work, commentaries on *see under* Aristotle
  attention given to earlier writings as key 87
  commentators adding own ideas/criticisms 93, 109
  communicating Greco-Roman ideas to other cultures 6
  different aims and motivations of commentators 88–92
  earlier texts, crafting in relationship to 6
  group activity, texts shared and read as part of 90
  important genre for communicating ideas 87
  lemmata 90–92, 109
  mathematics texts 87–88
  medieval period, importance in 88
  Meteorology, commentary on *see under* Aristotle
  nature of 20, 87–88
  non-classroom based audiences, writing for 90
  'open' character of genre of commentary 95, 99–100
  origins 6, 86
  pedagogical contexts, commentaries composed within 86, 88
  philosophical texts, commentaries on 89
    commentaries as philosophical works in own right 89
    commentary as significant genre for 87
  lemmata, organised by 90–92
  preservation of texts by commentators 88
  purposes/function of 86–87, 88–92, 109
  role of commentators in shaping how predecessors' works regarded 88
  school-texts, regarded as 86
  scientific topics, significant genre for writing about 20
  second-order status of 89
  teaching function of 89–90

commentary (*cont.*)
  textual issues, addressing 87
  transcriptions of lectures, as 89
Conics (Apollonius of Perga) 61, 62
Conon 63
Constantine Cephalas 39
Conte, Gian Biagio 5
Craik, Elizabeth 12
Croesus 43–44, 46–47, 48
counting
  Cattle Problem, in 32
  challenge of counting large numbers 34
  encouraging readers to count 34
    fruit and bowls, counting 32
  Greek Anthology, counting problems in 33, 34
  importance of 40
  isopsephic/equal pebbles used for counting 34
  perceptible numbers, concern of calculation with 45
  practical mathematics, as 41, 44, 48
  question-and-answer games featuring counting 26 *see also* numbers

de Anima (Aristotle) 93
De rerum natura see under Lucretius
Delatte, Armand 117, 122
Delian Apollo, hymn to 66–67
Demetrius
  (possible author of) *On Style* 50, 62
diagrams *see under* Aristotle
dialogue 1, 11
  authorial choice of, reasons for 3, 11
  few 'scientific' dialogues composed in ancient Greece and Rome 11
  gaining prominence in early modern period 2
  seldom used for scientific topics in antiquity 2
didactic poetry *see under* poetry
Diderot, Denis 73
Dillon, John 127
Diogenes Laertius
  Life of Epicurus 19, 50–51, 116
    account of Epicurus' life and works 116–17
    Epicurus promoted as ethical teacher 126
  Lives of Eminent Philosophers 13, 52–53, 115–16
    approach to biography 116–17
    intellectual lineage and relationships, describing 121
    Ionian succession 117–18, 121
    Italian succession 117–18, 121
    sources/shifting between sources 118, 120, 121, 123
    themes to be covered 117
    Pythagoras, *bios* of 21, 114, 116–21

accomplishments 119
author, activities as 118–19
education and travels 118
family 120
founder of philosophical succession 121
part of larger work, as 115–16
Pythagorean memoirs/notebooks, contents of 120
reincarnation, Pythagoras' views on 118, 119
teaching and lecturing 120
Dionysius of Halicarnassus
  letters to Ammaeus and Gnaeus Pompeius 68
Dionysodorus 50
  letters 71
Dirlmeier, Franz 15–17
Doody, Aude 73, 74–75, 77
Dositheus 63
Drabkin, I.E. 9, 32
Duff, David 6

education
  children 40–41
    education through play 40–41, 44
    letter-writing 63
  commentaries
    pedagogical contexts, commentaries composed within 86, 88
    school-texts, regarded as 86
    teaching function of 89–90
  didactic poetry, use of *see under* poetry
  enkuklios paideia
    educational aspiration of 74
    Greek 'system' of education in core subjects, as 73
    Sophists, and 73–74
Einstein, Albert 21
Elements (Euclid) 87–88
Empedocles 14
epic metre, use of 27, 28–29
encomia 21, 111–12
encyclopaedia 2, 11, 19, 72–85
  Natural History as encyclopaedia 72, 74, 75, 77–78 *see also under* Pliny the Elder
  label of 'encyclopaedia' 73
  meaning of 72–74
  new genre 131
    Natural History as important example of 77–78, 85
  Roman or imperial genre, as 77, 85
  Roman tradition adding practical lore to Greek rhetoric/philosophy 76
  scientific ideas and methods, as genre for communicating 77
  works dividing subjects into *artes*, and 74

Index

Encyclopédie (Diderot and le Rond d'Alembert) 73
Enk, Petrus Johannes 73–74, 77
Epicurean philosophy 30
  cosmology and meteorology 52–53, 54–55
  intention to communicate widely/alleviate anxiety 53–55
  nature 50–51
Epicurus
  letters 19, 52, 71
    didactic letters 52–53
    instructional nature of 19, 50–51
    natural philosophy, on 52–55
    scientific and mathematical topics, on 52
  Letter to Herodotus 52–54
  Letter to Pythocles 52–53, 54
  Life of Epicurus (Diogenes Laertius) 19, 50–51, 116
    account of Epicurus' life and works 116–17
    Epicurus promoted as ethical teacher 126
  natural philosophy, letters on 52–55
    audience, broad and literate nature of 53
    cosmology and meteorology 52–53, 54–55
    intention to communicate widely 53–55
  On Nature 29
  prose antecedents of poetry 29
  school of Epicurus/the Garden 53, 71 *see also* Epicurean philosophy
epigrammatic poetry *see under* poetry
epitomes 20, 87
Eratosthenes of Cyrene 19, 71
  Alexandrian Library, head of 56–57
  Archimedes, and
    competitor to 59
    corresponding with 35, 49, 51, 63, 64
    relationship between 38–39
  dedicatory letters 19
  geographical works, use of seals in 67–68
  Geography (*Geographica*) 38, 67–68
  Homer, interest in 67–68
  Letter to King Ptolemy 48, 55–59
    authenticity of 56, 61, 64
    choice of a letter 63
    criticisms of/challenges to authenticity 61, 64
    criticisms of, countering 61
    Delian problem 58, 59–60, 64, 65, 69, 71
    different audiences, operating for 60–61
    embellishments, use of 64
    epistolary styles 63–66
    Eratosthenes' reasons for writing 68–70
    heuristic text, as 68
    intellectual patronage, importance of 57
    instruments and tools in scientific work, use of 130
    layers of communication and meaning 59–62
    letters as gifts 52, 55–56, 70
    letter format providing 'envelope' for several genres/styles of writing 60, 69–70
    literary facets of 65
    monument to commemorate solution to problem 56, 59, 67, 68, 69
    operating on several levels 60
    poetry in 56, 60, 65–66
    preferred style of letter as plain 62
    problem of duplication of the cube, celebrating solution to 55–59
    reading of *Letter* aloud 66–67
    rich and complicated mathematical text, as 68
    signing and sealing 66–68
    sphragis, use of 67–68
    structures and genres 59–60
    technical terms and diagrams, use of 64
    text, nature of 57–59
    variety of styles and literary form 60, 64, 69–70, 144–47 *see also* text of letter (Appendix B)
  mathematical poetry 49
  number theory, interest in 39
  Platonicus 57
  scholar and poet, as 56–57, 64
  tutor, as 56–57
Euclid 20
  Elements 87–88
Eudoxus 29, 58
  Phaenomena 29
    Aratus' *Phaenomena* as later poetic version of 29
    Hipparchus' commentary on 88
eulogies 21, 111–12
Eutocius of Ascalon
  commentary on Archimedes' *Sphere and Cylinder* 55, 88
  Letter to King Ptolemy, and 55, 57

forgeries 70–71
  ease of forging 51–52
  value of forgeries 70–71
Fraser, P.M. 69
French, Roger 75

Gaius Plinius Secundus *see* Pliny the Elder
Gaius Scribonius Curio 63–64
Galen of Pergamum
  commentaries on Hippocratic works 88–89
  philosophical interests 88–89

Geminus
  introductory teaching texts 1
genres
  classificatory tool, as 4, 5
  commentary 6
  conventional nature of 5
  definitions of 4, 17, 72
  didactic poetry 6
  important bridge connecting authors and readers, as 133
  text 'traditions' as an abstraction 6
Geography (Geographica) (Eratosthenes of Cyrene) 38, 67–68
Gnaeus Pompeius 68
Goldhill, Simon 13
Greco-Roman, definition of 7
Greek Anthology (Anthologia Graeca) 18, 22, 69, 131
  authors, non-specialists as 47
  counting problems 33, 34
  mathematical problem-poems/epigrams (Book 14) 24, 33, 39–47
    apples, problems using 41, 43, 44
    children, education of 40–41
    collected by Metrodorus 39
    logistic, problems of calculation and 44–46
    myths and legends, and 169, 42–44
    poetry, uses of 49
    practical mathematics, different types of 41
    soluble nature of 40
    Thymaridas, rule of 46, 135–43 *see also* Arithmetical Epigrams (Appendix A)
  riddles and oracles also included 24–25, 26
  sources of 39
  Sphinx's Riddle 25, 48
  valued as poetry and as mathematics 24
Gregory of Nazianzus 64

Hadas, Moses 113–14, 127, 128
handbooks 11, 20, 87
Heavens, The (Cleomedes) 132
Heraclides of Pontus 118
Heraclitus 118
Herodotus 43, 48
heroes
  aretalogical accounts of heroic teachers as moral exemplars 113–14, 127
  divinity of heroic teacher, promoting 128
  'hero', meaning of 113, 114
  heroes, types of honours offered to 113
  heroic character of individuals, *bioi* emphasising 112–14
Herschbell, Jackson 127
Hesiod 3, 33
  agriculture

  astronomy, and 80, 81–82
  value of agricultural life 83
  authority on science, quoted as an 23
  canonical status of Hesiodic poems 12, 18, 28
  dating of poems 27
  great thinker, regarded as 27
  little known about 27
  metre, poems in 18
    epic metre 27, 28
    hexameter, use of 26, 28
  nature of poems 28
  question-and-answer games 26
  Theogony 15, 22, 42
    concerned with physical world 28
    poetic seal in 66
  Works and Days
    first example of didactic poem, as 23
    technical matters, referring to 28
Hipparchus 29
  commentary on *Phaenomena of Aratus and Eudoxus* 88
Hippocrates of Cos
  Galen's commentaries on Hippocratic works 88–89
  works attributed to 12
Hippocrates of Chios 58
Hippocratic Corpus 12
  Aphorisms 12
Historia Naturalis (Natural History) see under encyclopaedia; Pliny the Elder
History of Animals (Historia Animalium) (Aristotle) 100–1
Homer 33, 96
  authority on science, quoted as an 23
  canonical status of Homeric poems 12, 28
  dating of poems 27
  great thinker, regarded as 27
  little known about 27
  metre, poems in 18
    epic metre 27
    hexameter, use of 26
  nature of poems 28
  Odyssey 38
  oral 'tradition', part of 13
  question-and-answer games 26
Homeric Enquiries (Porphyry) 115–16
Høyrup, Jens 47, 49
Huffman, Carl 62, 65, 69
Hypatia of Alexandria
  commentaries 89

Iamblichus 46
  Christian threat to pagan philosophers, recognition of 126
  founding own school 116
  Neopythagorean or Neoplatonist, as 121

Index

On General Mathematical Science XXV 101–2
On Pythagoreanism 115–16, 123
  books of 124
  loss of books 124, 127–28
  mathematics, importance of studying 127–28
  pedagogical progression from general to more difficult 127–28
On the Pythagorean Life 21, 111, 114, 123–29
  accomplishments of Pythagoras 126
  character of Pythagoras 125
  competing with Christian gospels 121, 123
  divine guide, Pythagoras as 124
  divine lineage of Pythagoras 125
  education of Pythagoras 125
  exhortation to adopt Pythagorean philosophy 124–25, 127
  influence of Christianity in shaping *bioi* of Pythagoras 121
  part of larger work, as 115–16, 123
  philosophy dependent on link to the divine 124
  preliminary guide to Pythagorean philosophy, as 123
  Pythagoras' 'school', followers and members of 123, 126
  Pythagorean women 126
  teachings and advice 126
Introduction to Arithmetic (Nicomachus) 46
introductory texts/introductions 1, 11, 132
  continuing to be used 2
Isocrates 23

John Philoponus
  Christianity, and 93, 108
  commentaries 88, 89–90
    Aristotle's *Analytica*, on 93
    Aristotle's *Categories*, on 93
    Aristotle's *de Anima*, on 93
    Aristotle's *Metaphysics*, on 93
    Aristotle's *Meteorology*, on 92, 93, 96–98, 100–10
    Aristotle's *On Generation and Corruption*, on 93
    Aristotle's *Physics*, on 90–92
    divided into portions and delivered orally 90
    double commentary, providing 91–92
    thought experiment (ants tracking along a path) 106
    transcriptions of lectures, as 89
  teacher, as 93
Julius Caesar 81
  farmer's calendar 84

Kahn, Charles 10–11, 13, 14–15, 116, 121, 123
Kearns, Emily 113

Kepler, Johannes 114
Knorr, Wilbur 61, 65

Laws see under Plato
Learned Banqueters (*Deipnosophists*) (Athenaeus of Naucratis) 25–26
lectures 1, 132
Leon the tyrant of Phlius 118
Leonides of Alexandria 34
Lessing, Gotthold Ephraim 37
letter 11, 18–19, 50–71
  authenticity of/ease of forging 51–52, 70–71
  authors and addressees, relationships between 51, 131–32
  Christians, letters as important form of communication for 51, 52, 70
  conversations, as 70
  dedicatory letters 19
  delivery of letters 51, 68–69
  development of letter writing 51, 63
  Epicurus' letters *see under* Epicurus
  Eratosthenes' letters *see under* Eratosthenes of Cyrene
  famous individuals, letters attributed to 51
  features of well-composed letters/'style manual' 64
  forgeries, value of 70–71
  gifts, letters as 52, 55–56, 57, 70
  membership of specific communities, reinforcing 51, 70
  papyrus, use of 51, 68–69
  prefaces as type of letter 63
  preferred style of letter as plain 62
  public and private letters 18–19, 50–51
  purposes of letters 19, 50–51, 52, 63, 70
  scientific and mathematical topics, letters on 51, 52, 63, 68, 131–32
  sealing of letters using wax or lead 66
  technical mathematical texts, as preface to 63
  various shapes of letters 63
  well-written letters conforming to particular style 63–64
Letter to Herodotus see under Epicurus
Letter to King Ptolemy see under Eratosthenes of Cyrene
Letter to Pythocles see under Epicurus
Library, The (Apollodorus) 25
Life of Apollonius (Philostratus) 112–13
Life of Epicurus see under Diogenes Laertius
Life of Pythagoras see under Porphyry
literary formats, genres and types of texts 4–6
  authors' aims 6
  genres, definitions of 4
Lives of Eminent Philosophers see under Diogenes Laertius

Index

Lloyd, G.E.R. 8, 78
Lucretius
  On the nature of things (De rerum natura)
    18, 74–75, 131
  prose antecedents of 29
  didactic poetry 24, 31
  making subjects more enjoyable to readers
    30, 132
  epic metre, use of 27

Manilius
  Astronomica 24, 30
Marcus Tullius Cicero 7
mathematics, writings on
  commentaries on mathematical texts
    87–88, 89
  letter 51, 63, 68, 131–32
  mathematical epigrams 18, 24, 31–32
    Cattle Problem see Cattle Problem
      (*problema bovinum*)
    Greek Anthology see Greek Anthology
      (*Anthologia Graeca*)
    intellectual engagement, encouraging 24
    myth and epic, allusions to 33, 42–44
    poetic traditions, reflecting 48
    poetry, uses of 49
    posing problems to be solved 31–32
    practical problems, riddles about life
      and 48–49
    question-and-answer games 26
    riddles, as 24, 48, 49 *see also* riddles
    role in wider Greek culture 24–25, 26
    valued as poetry and as mathematics 24
    verse, often presented in 26
    ways in which audiences encountered
      mathematical epigrams 48–49
  mathematics as poetry/mathematical poems
    30, 32–35
    authors 32
    epigrams 31
    forgeries 34–35
    mathematical problem-poems 33, 34
    metre 32
  Pythagoras, and *see* Pythagoras
  range of literary formats adopted 1–2,
    32, 131–32
  specific character/key features of Greek
    mathematical texts 64 *see also* science
    writings
Maximus Planudes 39
medical writings 2
  Hippocrates
    Galen's commentaries on Hippocratic
      works 88–89
    works attributed to 12

Hippocratic Corpus 12
letter 51, 68
Menaechmus 58
meteorology
  Epicurean philosophy 52–53, 54–55
  Meteorology (Aristotle) *see under* Aristotle
  Natural History, in 77, 78–85
    animal behaviour and plants, importance
      of 83, 85
    astrometeorological calendars, questioning
      usefulness of 83
    astronomy 80–81, 82–83, 85
    causes of meteorological phenomena,
      accounts of 80
    convergence and culmination of Greek and
      Roman traditions, as 78
    explaining and predicting meteorology
      78, 79, 85
    discriminating in the selection of
      material 79–80
    explanation as part of natural
      philosophy 79
    farmers' almanacs/calendars 81–82, 83, 84
    farming/agriculture 79, 80–83, 85
    imperial endeavour, meteorology to be
      undertaken as 79–80
    observation, importance of 78–79,
      82–83, 85
    prediction as part of agricultural
      skill 79
    reliance on work of others 78–79, 80
    unification of diverse meteorological
      traditions, as 84–85
    weather prediction 80–81, 84, 85
  parallel traditions of meteorological
    prediction and explanation 78
  reliance on work of predecessors as key
    characteristic 78
metre *see under* poetry
Metrodorus of Tralles 33, 39, 40
Mithradates VI, King 34–35
Most, Glenn 5
Mourelatos, Alexander P.D. 29
Murphy, Trevor 73, 74–75, 76–77

Natural History see under encyclopaedia; Pliny
  the Elder
Netz, Reviel 37, 55
Newton, Isaac 9–10
Nicobulus 64
Nicomachus
  Introduction to Arithmetic 46
numbers
  abstract idea, as 44
  bowl numbers 44–46

Index

Cattle Problem, very large number as solution to 37
figured numbers 37
harmonics study based on 101
intellectual bonds through numbers and poetry 39
isopsephic poems 34
large numbers, fascination with 34
number theory 41
   Eratosthenes' interest in 39
   logistic, and 41, 44–46
   question-and-answer games featuring numbers 26
   Riddle of the Sphinx, numbers involved in 25–26, 48
   sheep-numbers 45 *see also* counting

Odyssey (Homer) 38
Oedipus 25
Olympiodorus
   commentaries
     Aristotle's *Meteorology*, on 92, 93, 98–99, 100–10
     divided into lessons 90
     double commentary, providing 91–92
     teaching function of 90
     transcriptions of lectures, as 89
     works of Plato and Aristotle, on 86, 92, 93
   teacher, as 93
O'Meara, Dominic 121
On Agriculture (Columella) 23
On Education (Pythagoras) 118
On General Mathematical Science XXV (Iamblichus) 101–2
On Generation and Corruption (Aristotle) 93
On the heavens (Aristotle) 88
On the Life of Plotinus (Porphyry) 86, 90
On Nature (Epicurus) 29
On Nature (Parmenides of Elea) 28–29
On Nature (Pythagoras) 117–18
On the nature of things see under Lucretius
On the Pythagorean Life see under Iamblichus
On Pythagoreanism see under Iamblichus
On Riddles (*Peri griphōn*) (Clearchus of Soli) 26
On the Sphere and the Cylinder see under Archimedes
On Statesmanship (Pythagoras) 118
On Style (Demetrius) 50, 62

Palatine Anthology (Constantine Cephalas) 39
Parallel Lives (Plutarch) 112
Parmenides of Elea 14
   epic metre, use of 27, 28–29
   On Nature 28–29

preservation of fragments of texts by Simplicius 88
pebbles
   figured numbers, creating 37, 37n. 31
   isopsephic/equal pebbles used for counting 34
Pelling, Christopher 111–12
Peri physeōs (*On nature*) (Anaximander) 10–11, 13
Pfeiffer, Rudolf 65
Phaenomena see under Aratus of Soli
Pherecydes 117
Philosophical History (Porphyry) 115–16, 121
Philostratus
   Life of Apollonius 112–13
Physics see under Aristotle
Planudean Anthology (Maximus Planudes) 39
Plato
   Aristoxenus' biography of 112
   arithmetic and logistic, distinction between 41
   canonical author, as 18
   Charmides 44–46
   commentaries on works of 86
   target texts, Plato's works as 87
   dialogues 3, 11, 16
   Laws
     educating children through play 40–41, 44
     logistical problems 45–46
     making subjects more enjoyable to readers 30
   letters attributed to Plato regarded as spurious 51
   Pythagorean philosophy, Plato creating intellectual heritage of 115
   Republic 40
     practical mathematics, different types of 41
     style of Greek scientific explanation 9–10
   Timaeus
     Calcidius' commentary on 88
     Porphyry's commentary on 115–16
Platonicus (Eratosthenes of Cyrene) 57
Pliny the Elder
   astronomy in *Natural History*
     astronomical knowledge, insufficiency of 83, 85
     astrometeorological calendars, questioning usefulness of 83
     difficulties of 81
     farming, and 80–81, 82–83
     trade and navigation, and 80–81
     weather prediction 84
   career 76
   Dionysodorus, on 71
   Hesiod as authority on astronomy 28
   Natural History (*Historia Naturalis*) 19, 50, 72
     agriculture 79, 80–83, 85
     encyclopaedia, as 72, 74, 75, 77–78, 85

Pliny the Elder (*cont.*)
  enkuklios paideia, Pliny's concept of 73
  function of 75
  genre 74, 75–78, 85
  imperial context and aims 75–77, 85
  importance of 72, 73, 75–77
  interconnectedness between different bodies of knowledge, demonstrating 85
  meteorology 77, 78–85
  learned communication as new 'genre', development of 77
  'mere' compiler of theories/opinions of others, Pliny as 85
  nature as principal theme/meaning of 'nature' 74, 75
  precedent for later works 73
  preface as free-standing epistolary text 63
  Roman work/concerned with usefulness and order, as 75–77
  scientific ideas and methods, communicating 77
  structure and organisation 74–75
  summarium 74, 75, 78
  summary of contemporary knowledge and different points of view 77
  Titus, addressed to 76–77
  meteorology in *Natural History* 77, 78–85
    animal behaviour and plants, importance of 83, 85
    astrometeorological calendars, questioning usefulness of 83
    astronomy 80–81, 82–83, 85
    causes of meteorological phenomena, accounts of 80
    convergence and culmination of Greek and Roman traditions, as 78
    discriminating in selection of material 79–80
    explaining and predicting meteorology 78, 79, 85
    explanation as part of natural philosophy 79
    farmers' almanacs/calendars 81–82, 83, 84
    farming/agriculture 79, 80–83, 85
    imperial endeavour, meteorology to be undertaken as 79–80
    observation, importance of 78–79, 82–83, 85
    prediction as part of agricultural skill 79
    reliance on work of others 78–79, 80
    unification of diverse meteorological traditions, as 84–85
    weather prediction 80–81, 84, 85

Plotinus 20
On the Life of Plotinus (Porphyry) 86, 90
Plutarch 57
  ethical and moral matters in biography, interest in 112
  dialogue form 1
  Parallel Lives 112
poetry 1, 17–18, 22–49
  authorial choice of, reasons for 3, 132
  commentaries, subjects of 20
  didactic poetry 30
    educational/ instructional/pedagogical function 23, 24
    entertainment, for 24
    free-standing text, as 24
    hexameter, use of 23–24
    lengthy nature of 24
    making subjects more enjoyable to readers 30
    nature of 6
    scientific and technical ideas, conveying 23–24
    whether a distinct genre 23–24
  discarded in scientific contexts 2
  earliest surviving Greek works, as 22–23
  early epic poems: traditional and authoritative 27–29
  epigrammatic poetry 24–26, 30–32
    active reading, encouraging 31
    brief/concise nature of 5, 30, 31
    commemorating individuals and events 30
    entertainment, for 30
    epic poetry, and 31
    free-standing text, as 24
    functions of 31
    material used 30
    mathematical epigrams *see under* mathematics, writings on
    myth and epic, allusions to 42–44
    not read only as individual texts 24–25
    occasional poetry, as 30
    performed or recited 30
    riddles, oracles and mathematical problems, and 31
    salient features of 31
  isopsephic poems 34
  mathematics as poetry/mathematical poems 30, 32–35
    authors 32
  metre 32
    cultural values, and 6
    dactylic hexameter/epic metre 26, 27
    didactic poetry 6, 23–24
    epic poetry, use of hexameter in 26, 27
    hexameter, use of 23–24, 26, 27

Homeric and Hesiodic poems 18
  importance in archaic period 12–13
  poetry understood as being in
    metre 22–23
  powerful medium for
    communication, as 28
  riddles 26
  text linking to another text through
    metre 6
  oral composition/performance, written
    preservation secondary to 13
  power/authority of poetry in Greco-Roman
    world 12, 13, 18
  powerful means of communicating 22, 28
  prose to poetry, from 29
  quotations of poetry within prose text as
    authoritative statements 23
  riddles *see* riddles
  scientific discourse, important form
    of 18
  'seal' (*sphragis*) as closing to poetry, literary
    use of 66–68
  sphragis allowing identification of the
    poet 67
  singing as well as reading poetry 22
  treatise, as 14–15
  valuing poetry over prose formats 23
Porphyry
  Against the Christians 126
  breadth/scope of works 115–16
  Christian threat to pagan philosophers,
    recognition of 126
  commentaries 20, 90
    Physics (Aristotle), on 115–16
    Timaeus (Plato), on 115–16
  Homeric Enquiries 115–16
  Life of Pythagoras 21, 114, 121–23
    character of Pythagoras 122
    daily life 122–23
    diet/views on diet 122–23
    education 122
    family and origins 122
    influence of Christianity in shaping *bioi* of
      Pythagoras 121
    part of larger work, as 115–16, 121
    school, founding 122
    sources 122
    travels 122
  Neopythagorean or Neoplatonist, as 121
  On the Life of Plotinus 86, 90
  Philosophical History 115–16, 121
Posterior Analytics (Aristotle) 9–10
Praechter, Karl 90
Proclus 44–46
  commentary on Euclid 20, 41

Elements 87–88
mathematical poems 41
prose works 1, 10–15
  aphoristic works 12
  authoritative expression, prose as
    medium for 13
  commentaries, subjects of 20
  earliest writers of Greek prose 12–13
  formats of prose works 11
  'Ionian prose treatise' as new form of prose
    literature 12
  importance of
    communicating scientific ideas, importance
      of prose for 13, 14
    scientific medical and technical writings,
      in 10–11
  medical writings among earliest prose
    works 12
  pragmateia, meaning of 15–17
  prose to poetry, from 29
  quotations of poetry within prose text as
    authoritative statements 23
  reasons for use of prose 12–13
  use of prose as significant cultural
    shift 12–13
  technical prose writing, tradition of 11
Ptolemy (Claudius Ptolemy)
  mathematics
    branch of theoretical knowledge, as 8
    specialist mathematical texts subject of
      commentaries 89
  prose works 1
Ptolemy I, King 57
Ptolemy II, King 57
Ptolemy III, King 55, 56–57
  Letter to King Ptolemy see under Eratosthenes
    of Cyrene
Pythagoras
  accomplishments 119, 126
  author, activities as 118–19
  biographies of
    Aristoxenus 112, 116, 119
    Diogenes Laertius 21, 114, 116–21
    Iamblichus' *On the Pythagorean Life* 21, 111,
      114, 123–29
    Porphyry's *Life of Pythagoras* 21, 114, 121–23
  character 122, 125
  daily life 122–23
  diagrams, use by Pythagoreans of 101–2
  diet/views on diet 119, 122–23
  divinity of 121, 123, 125
  divine guide, Pythagoras as 124
  divine lineage 125
  education and travels 118, 122, 125
  views on education 120, 126

Pythagoras (*cont.*)
  family and origins 120, 122
    divine lineage 125
  founder of philosophical succession 121
  heroic status of 114
  mathematics
    importance of 128
    mathematician, Pythagoras as 119
    Pythagorean mathematics 127–28
  On Education 118
  On Nature 117–18
  On Statesmanship 118
  physicist, as 117–18
  prayer, views on 118–19
  Pythagoreanism
    akousmatikoi 115
    Plato creating intellectual heritage 115
    mathematikoi 115
    religious/ethical or sectarian form 115
    scientific/mathematical or philosophical form 115
  reincarnation, views on 118, 119
  school, Pythagoras'
    development of teaching methods 123
    followers/members of 123
    founding 122
    longevity of 120
  teaching and lecturing 120, 123, 126
  temperance 118–19
  Thales, student of 125
  widely revered 114

question-and-answer texts/games 26
  continuing to be used 2

Rabelais, François 73
Republic see under Plato
riddles 25–26
  epigrams, and 31
  mathematical epigrams 24, 48, 49
  mathematical practitioners using 49
  not read only as individual texts 24–25
  oracular questions, kinship with 26
  posed in various contexts 24–25
  principal collections of extant Greek riddles 26
  question-and-answer games 26
  riddle of the Sphinx 25–26
  verse, often presented in 26
Roberts, Deborah 66
Rond d'Alembert, Jean le 73
Rosenmeyer, Patricia 51

Sand-Reckoner see under Archimedes
'science', 'scientific' and 'scientist' 7–9
  biographies of scientists *see* biography
  knowledge, nature of 8–9
  natural philosophy, meaning of 8–9
  'scientific', use of term 7–9
  'scientist', meaning of 7
science writings
  historiographical context 3–4
  literary formats
    authorial choices, reasons for 2–3, 6, 132
    changes in 2
    commentary 20
    encyclopaedia *see* encyclopaedia
    letter 51, 52, 68 *see also* letter
    new literary forms 2
    poetry 18, *see also* poetry
    prose works *see* prose works
    range of 1–2, 130–31
  long 'afterlife' of 133
  reasons for studying 133
  technical writings
    interest in studying as texts 4
    not regarded as 'literary' 4
  texts, describing 9–17
    aphoristic works 12
    authors, types of 10
    'axiomatic-deductive system', influence of 9–10
    categories not clear-cut 9
    Greek terminology and relationship to modern descriptions 10–17
    pragmateia, meaning of 15–17
    scientific ideas, importance of prose for communicating 13
    treatises/prose texts 10–15
    use of prose, reasons for 12–13
    wide variety of types of texts 10 *see also* prose works; treatises
  writing and reading as part of broader Greco-Roman culture 2 *see also* mathematics, writings on; medical writings; technical writings
Seneca
  letters 70
  Moral Epistles 70
sheep (*mēlon*)
  'apple', and 45–46
  calculational problems, use in 45–46
  sheep numbers 45
Simplicius
  commentary on Aristotle's works
    Categories 87
    non-classroom based audiences, writing for 90
    On the heavens 88
    Physics 88

Index

preservation of fragments of Parmenides'
  texts 88
  written for publication 89
Sluiter, Ineke 89, 90
Smith, Morton 113–14, 127, 128
Socrates
  Aristoxenus' biography of 112
  Diogenes Laertius' biography of 117
Solon 48
Sophists 73–74
Sosicrates 118
Sotion of Alexandria 112
Soutar, George 27
Sphinx's Riddle 25–26, 48
Strabo 67–68

Taavitsainen, Irma 5
technical writings
  communicated in range of formats 2
  not regarded as 'literary' 4
  notes and memoranda 11, 12
  prose, and
    communicating scientific ideas, importance of prose for 13, 14
    'Ionian prose treatise' as new form of prose literature 12
    scientific medical and technical writings, importance of prose in 10–11
    tradition of technical prose writing 11
Thales of Miletus
  business acumen 111
  Diogenes Laertius' biography of 116, 117
  first natural philosopher, as 13, 111
  intellectual unworldliness, claim to 111
  Pythagoras as student of 125
  whether author of any works 13
Theogony see under Hesiod
Theon of Alexandria
  commentaries 1
  mathematical texts, on 89
Theon of Smyrna 57
Theophrastus 15, 80
Thesleff, Holger 66–67
Thymaridas 46
Timaeus see under Plato
Titus 63
Totelin, Laurence 34–35

Trapp, Michael 68
treatises 6, 10–15
  first Greek prose treatise 10–11, 13
  'Ionian prose treatise' 12
  modern English use of term 14, 15–16
  peri physeōs (*On nature*) treatises 14–15
  functional definition of 15
  prose treatise for communicating scientific ideas, importance of 13, 14
  'treatise'
    general term for writing on specific subject, as 14
    no characteristic form defining 'treatise' 14
    not needing to be a prose work 14–15
    usefulness of referring to works as treatises unclear 15
    written 'work' rather than particular genre, treatise as 11, 14

van der Eijk, Philip 11
Varro
  encyclopaedic work/subjects divided into *artes* 74
Virgil 81
  agricultural life, value of 83
  epic metre, use of 27

Wallace-Hadrill, Andrew 75
Werlich, Egon 5
Whewell, William 7
Whitehead, Alfred North 114
Wilamowitz-Moellendorff, Ulrich von 61, 64, 69
women
  Dionysodorus' letter, first readers of 71
  Epicurus' school, members of 53, 71
  female literacy 71
  Pythagorean women 126
Works and Days see under Hesiod

Xenophanes 14

Yatsuhashi, Akira 31

Zeller, Eduard 123